AF601239

AN INTRODUCTION TO
HORTICULTURE

The Author

Dr Praveen Dhar T. [M. Sc., M. Phil (Botany) M. Ed., M. Phil (Education), M.A. (History) and Ph.D. (Botany)] is presently Assistant Professor in the Department of Botany at St. Stephen's College, Pathanapuram, Kollam, Kerala. Dr. Praveen has more than two decades of teaching and research experience. He has published eight books and more than 80 research papers in national and international journals of repute. He has also presented various research papers in several national and international conferences. His area of interests are botany, forestry, horticulture, mushroom cultivation, education, research methodology and environmental studies. He also acting as an editorial board member of research journals.

AN INTRODUCTION TO
HORTICULTURE

Dr Praveen Dhar T.
Assistant Professor
Department of Botany
St. Stephen's College
Pathanapuram, Kollam,
Kerala

2020
Daya Publishing House®
A Division of
Astral International Pvt. Ltd.
New Delhi – 110 002

ISBN: 9789390384266 (Int. Edition)

Published by : **Daya Publishing House®**
A Division of
Astral International Pvt. Ltd.
– ISO 9001:2015 Certified Company –
4736/23, Ansari Road, Darya Ganj
New Delhi-110 002
Ph. 011-43549197, 23278134
E-mail: info@astralint.com
Website: www.astralint.com

PREFACE

Horticulture is a science of studying garden plants. The word Horticulture is derived from two Latin words *viz. Hortus* means garden and Culture means knowledge of growing crops. Horticulture is an aesthetic science that deals with the important crops which are grown in the gardens *e.g.* vegetable crops in vegetable garden, fruit crops in fruit orchards. India with diverse soil and climate comprising several agro-ecological regions provides ample opportunity to grow a variety of horticulture crops.

Horticultural crops form a significant part of total agricultural produce in the country comprising of fruits, vegetables, root and tuber crops, flowers, ornamental plants, medicinal and aromatic plants, spices, condiments, plantation crops and mushrooms. However, per capita consumption of fruits and vegetables in India is only around 46kg and 130g against a minimum of about 92g and 300g respectively recommended by Indian Council of Medical Research and National Institute of Nutrition, Hyderabad. With the present level of population, the annual requirement of fruits and vegetables will be of the order of 32.58million tonnes and 83million tonnes respectively. To meet this requirement the National Commission on Agriculture has projected an area of 4m.ha. and 8m.ha. Unlike other text books, which largely focus on the one horticultural aspects, present text book providing a uniquely balancing concepts, contains missing concepts from other text books in the area and topics corresponding college course. These topics are ideal for post graduates and undergraduates' introductory courses contain numerous examples for students to nourish their needs.

I believe that this book will be also useful for fellow teachers, researchers and development officers for references and easy answering of many complicated questions.

I am grateful to all those persons as well as various books, manuals, periodicals, magazines, newsletters, journals etc. that helped in the preparation of this book.

In spite of the best efforts, it is possible that some errors may have occurred into the compilation and editing of the book. Further queries, suggestions and criticisms for improvement of the book are always welcome and shall be thankfully acknowledged. Last but not the least, it is a pleasure for us to extend our sincere thanks to Mr. Anil Mittal, Astral International Pvt. Ltd. and his team members for his keen interest shown in preparation and publishing of this book so efficiently and promptly.

Dr Praveen Dhar T.

CONTENTS

Chapter 1

HORTICULTURE

1.1 Introduction

The word horticulture comes from two Latin words which mean "garden" and "culture." Horticulture is the art and science of growing and handling fruits, nuts, vegetables, herbs, flowers, foliage plants, woody ornamentals, and turf. It is a science on the cutting edge of biotechnology, an art, profession, business, industry, hobby, way-of-life, and therapy for millions of people. Each of us comes in contact with horticultural products and professions every day of our life.

Horticulture is the branch of agriculture that deals with the art, science, technology, and business of growing plants. It includes the cultivation of medicinal plants, fruits, vegetables, nuts, seeds, herbs, sprouts, mushrooms, algae, flowers, seaweeds and non-food crops such as grass and ornamental trees and plants. It also includes plant conservation, landscape restoration, landscape and garden design, construction, and maintenance, and arboriculture. Horticulturists apply their knowledge, skills, and technologies used to grow intensively produced plants for human food and non-food uses and for personal or social needs. The study and science of horticulture dates all the way back to the times of Cyrus the Great of ancient Persia, and has been going on ever since, with present-day horticulturists such as Freeman S. Howlett and Luther Burbank.

Horticulture is the area of agriculture involving the science of growing and caring for plants. Horticulture is socially important because it improves how we use plants, for food and other human purposes, as well as repairing the environment and personal aesthetics. Plants are also very important in environmental protection. Plants also have an important role in the beautification of urban and rural landscapes and recreation areas.

1.2 Divisions of Horticulture

There are five main branches of horticulture that are divided according to the type of crops produced and how the plants are used.

a. **Floriculture** – This area of horticulture focuses on the cultivation of flowers (cut and potted) and foliage. Flower arrangement also fall under this branch.

b. **Pomology** –This branch of horticulture revolves around production and cultivation of fruit crops.

c. **Nursery/Plant Propagation** – The development and dissemination of plant seeds, shrubs, trees, ornamental plants, and ground covering is the focus of this area of horticulture.

d. **Olericulture** –Olericulture handle the farming, processing, storage, and marketing of all edible parts of vegetables including the roots, leaves, flowers, stems, seeds, and young parts.

e. **Landscape Horticulture** –Landscape horticulture deals design, construct, and take care of landscapes in homes, businesses, and public areas.

Other Branches are

1. **Plantation Crops**: Refers to cultivation of crops like coconut, arecanut, rubber, coffee *etc.*
2. **Spices Crops:** Refers to cultivation of crops like, cardamom, pepper, nutmeg *etc.*
3. **Medicinal and Aromatic Crops:** Deals with cultivation of medicinal and aromatic crops.
4. **Post-Harvest Technology**: Deals with post-harvest handling, grading, packaging, storage, processing, value addition, marketing *etc.* of horticulture crops.
5. **Plant Propagation:** Deals with propagation of plants.

1.3 Importance and Scope of Horticulture

Importance and scope of horticulture are as follows:

1. Yield is high for per unit area: As compared to the field crops per hectare yield of horticulture crops is very high.
2. High returns per unit Area: From one unit area of land more income will be obtained compared to other crops.
3. A free grower/labour remains engaged for the whole: An opportunity for maintaining labours throughout the year like the cereals where one cannot keep himself and employ the labours during the slack season.

4. Best utilization of waste land: Some fruit crops can offer best utilization of waste land crops like wood apple, custard apple, karonda, litchi *etc.* can be grown in such areas.
5. Raw material for industries: Fruit farming is the base for several industries like canning, essential oils *etc.* which in turn provide work for more people.
6. Use of undulating lands: Fruit growing can be practiced in places where the gradient is uneven or where the land is undulating and agronomical crops cannot be cultivated.
7. Fruits and vegetables are the important energy giving material to the human body.
8. As compared to field crops Horticultural crops give more returns per unit area (More yield in terms of weight and money).
9. Horticulture crops are important as their nutritional status is high. Particularly fruits and vegetables provide high amount of vitamins and minerals to us.
10. Horticulture is important as it beautifies the surroundings.
11. Horticulture crops are suitable for small and marginal farmers.
12. The varieties of crops are available in the Horticulture section with wide range of uses.
13. Horticultural plants improve environment by reducing pollution, conserves soil and water and improve socio-economic status of the farmer. Scope of Horticultural crops.

1.4 Principles of Garden Making

Garden design is the art and process of designing and creating plans for layout and planting of gardens and landscapes. Elements of garden design include the layout of hard landscape, such as paths, walls, water features, sitting areas and decking; as well as the plants themselves, with consideration for their horticultural requirements, their season-to-season appearance, lifespan, growth habit, size, speed of growth, and combinations with other plants and landscape features.

The three categories below contain the basic elements that constitute the generally accepted version of good garden design.

a. Order/Balance/Proportion

This refers to the basic structure of the garden. Order can be obtained through symmetry, as in a formal garden, through repetition of plants or colors, or through balancing bold or bright features with a complementary weight of fine texture or muted features.

b. Harmony/Unity

Harmony is achieved in editing. This can be accomplished by using a limited color palette, the repetition of key plants throughout the garden.

c. Flow/Transition/Rhythm

Great structure and unifying elements will create a pretty picture, but your garden can appear static.

Other Principles are:

1. **Unity** should be one of your main goals in your design. It may be better understood and applied as consistency and repetition. Repetition creates unity by repeating alike elements like plants, plant groups or decor throughout the landscape.
2. **Simplicity** is actually one of the principles in design and art. Simplicity in planting, for instance, would be to pick two or three colors and repeat them throughout the garden or landscape.
3. **Balance** in design is just as the word implies, a sense of equality. There are basically two types of balance in landscape design these are, Symmetrical and Asymmetrical. Symmetrical balance is where there are more or less equally spaced matching elements of the garden design. Asymmetrical balance is not dependent on the shape of the garden.
4. **Color** adds the dimension of real life and interest to the landscape. Colors can also be used to direct your attention to a specific area of the garden. A bright display among cooler colors would naturally catch the eye.
5. **Natural transition** can be applied to avoid radical or abrupt changes in your landscape design. Transition is basically gradual change. It can best be illustrated in terms of plant height or color.
6. **Line** is of the more structural principles of landscape design. It can mostly be related to the way beds, walkways and entryways move and flow. Straight lines are forceful and direct while curvy lines have a more natural, gentle, flowing effect.
7. **Proportion** simply refers to the size of elements in relation to each other. Of all the principles of landscape design, this one is quite obvious but still requires a little thought and planning. Most of the elements in landscape design can be intentionally planned to meet the proper proportions.
8. **Repetition** is directly related to unity. It's good to have a variety of elements and forms in the garden but repeating these elements gives variety expression.
9. **Unity** is achieved by repeating objects or elements that are alike. Too many unrelated objects can make the garden look cluttered and unplanned.

1.5 Types of Pots and Containers

There are several types of containers that can be used for growing vegetables including polyethylene plastic bags, clay pots, plastic pots, metallic pots, milk jugs, ice cream containers, bushel baskets, barrels, and planter boxes. It is important to use containers that can accommodate roots of the vegetables you want to grow as the vegetables vary in sizes and rooting depths. The container needs to have good drainage, and should not contain chemicals that are toxic to plants and human beings.

The different types of containers are given below:

a. Pressed Paper Container

Pressed paper containers are a great choice which are made up of thick paper. These containers can breathe well, promoting healthy root growth and improving aeration, as well as insulating the roots from temperature changes that might otherwise harm or stress your plants. These are biodegradable, but will need to replace them every single year. Since the cost of these is low, the pots are still an economically sound choice. Some pressed paper pots are now being lined with a wax coating, which gives them a slightly longer life span.

b. Coir Container

Coir containers are economically and ecologically sound choices. These are made from coconut husks and are durable than the pressed paper pots, but still keep the benefits for the paper pots. Other types are made from grain husks and various binding agents rather than coconut husks. These pots are inexpensive and can be found in a wide variety of colors and shapes, up to around a foot in diameter.

c. Ceramic Container

Ceramic containers or stoneware, are made from finely textured, light-colored clay and then glazed. These containers are fired at a high kiln temperature, which reduces the pot's porousness and vulnerability to the elements. Ceramic pots are also tend to be expensive, but are available in a wide variety of color glazes and designs. They come in a wide variety of shapes and sizes, with large bowls being among the most common.

d. Terracotta Container

Terracotta is a type of clay that is commonly used in making pots and planters of various sizes and shapes. Terracotta pots can be as small as two inches in diameter or height, and as large as the creator's imagination. The shapes and sizes of containers from this material are seemingly endless, so they are quite versatile. It is typically a warm reddish brown color and offers an earthy appeal to gardeners.

e. Fiberglass and Resin Container

These are a pots which are the blend of resin and glass fibers. Fiberglass are lightweight, durable, and look very much like the materials they are molded to imitate. These containers do not need any special storage, they can handle any expected weather without having to be taken indoors.

f. Plastic Container

These are container which are made up of plastic material contain polythene. Most plants that you purchase in a store or at a nursery will go home with you in a plastic container, this is because it is the cheapest container for commercial growers to use.

g. Wood Container

Wooden containers are mainly made up of hard wood of plants. These containers tend to be square or rectangle, although there are some curved containers available as well. Wood is not likely to crack in cold weather, and is slow to dry out. The only real danger with wood containers is there is a chance of degrade or rot.

h. Metal Container

The material for manufacturing metal containers are by using hard or light metal. Metal containers are extremely durable, and in the case of case iron, extremely heavy. Metals such as aluminum can offer gardeners the durability of metal with a lighter weight, making the planting and moving of pots much more practical. Aluminum does not rust, does not need painting and costs less than some other metal options. Other metal choices include copper-coated stainless steel, zinc, lead, and copper.

i. Concrete Container

Concrete containers can look great, especially once they have waged a little. While concrete is the practical choice for large plants that might need the ballast support against the wind because of its sheer weight, which weight also makes the planter difficult to move.

1.6 Potting Mixture and Potting Media: Soil, Sand, Peat, Sphagnum moss and Vermiculite

The material in which the plants grow is called the "growing medium". Dozens of different ingredients are used in varying combinations to create homemade or commercial growing media.

Functions of growing or potting medium are:

1. Supply roots with nutrients, air and water
2. Allow for maximum root growth
3. Physical support to the plants

The different potting or growing media are:

1. Soil

Soil is a heterogeneous material consisting of three major components, a solid phase, a liquid phase and a gaseous phase. All three phases specifically influence the supply of nutrients to plant roots. The solid phase is the main nutrient reservoir. The inorganic particles of this phase contain cationic nutrients such as K, Na, Ca, Mg, Fe, Mn, Zn, and Cu, while the organic particles form the main reserve of N and to a lesser extent also of P and S. The liquid phase of the soil, the soil solution, facilitates nutrient transport in the soil, *e.g.* for the transport of nutrients from various parts of the bulk soil to plant roots. The gaseous phase contain soil air.

2. Sand

Sand is the basic component of soil, ranges in particle size from 0.05mm to 2.0mm in diameter. Fine sands (0.05mm – 0.25mm) do little to improve the physical properties of a growing media and may result in reduced drainage and aeration. Sand is a valuable amendment for both potting and propagation media. Sand is a naturally occurring granular material composed of finely divided rock and mineral particles. It is defined by size, being finer than gravel and coarser than silt. Sand can also refer to a textural class of soil or soil type; *i.e.* a soil containing more than 85 per cent sand-sized particles by mass.

3. Peat

Peat moss is formed by the accumulation of plant materials in poorly drained areas. The type of plant material and degree of decomposition largely determine its value for use in a growing medium. Although the composition of different peat deposits vary widely, four distinct categories may be identified. Peat also called turf is an accumulation of partially decayed vegetation or organic matter that is unique to natural areas called peatlands, bogs, mires, moors, or muskegs.

4. *Sphagnum moss*

Sphagnum moss – is the dehydrated remains of acid-bog plants from the genus Sphagnum (*i.e.* Spapillosum). It is light in weight and has the ability to absorb 10 to 20 times its weight in water. This is attributed to the large groups of water holding cells, characteristic of the genus. Sphagnum moss is perhaps the most desirable form of organic matter for the preparation of growing media. Drainage and aeration are improved in heavier soils while moisture and nutrient retention are increased in lighter soils. Sphagnum is a genus of approximately 380 accepted species of mosses, commonly known as peat moss.

5. Vermiculite

Vermiculite is a hydrated magnesium aluminum silicate mineral which resembles mica in appearance. It is found in various parts of the world including Australia, Brazil, Bulgaria, Kenya, Russia, South Africa, Uganda, USA and Zimbabwe.

Vermiculite is mined using open cast mining techniques where the ore is separated from other minerals and then screened or classified into several basic particle sizes.

1.7 Soil Types and Soil Preparation

Soil science has two basic branches of study - edaphology and pedology. Edaphology is concerned with the influence of soils on living things. Pedology is focused on the formation, description (morphology), and classification of soils in their natural environment. In terms of soil texture, soil type usually refers to the different sizes of mineral particles in a particular sample. Soil is made up in part of finely ground rock particles, grouped according to size as sand and silt in addition to clay, organic material such as decomposed plant matter. The ratio of these sizes determines soil type: *clay*, *loam*, *clay-loam*, *silt-loam*, and so on.

The different soil types are given below:

1. Sandy

Sandy soil has the largest particles among the different soil types. Water drains rapidly, straight through to places where the roots, particularly those of seedlings, cannot reach. Plants don't have a chance of using the nutrients in sandy soil more efficiently as they're swiftly carried away by the runoff. The upside to sandy soil is that it's light to work with and warms much more quickly in the spring.

2. Silty

Silty soil has much smaller particles than sandy soil so it's smooth to touch. When moistened, it's soapy slick. Silty soil retains water longer, but it can't hold on to as much nutrients as you want. Due to its moisture-retentive quality, silty soil is cold and drains poorly. Silty soil can also easily compact, so avoid trampling on it when working in garden. It can become poorly aerated, too.

3. Clay

Clay soil has the smallest particles and has good water storage quality. It's sticky to the touch when wet, but smooth when dry. Due to the tiny size of its particles and its tendency to settle together, little air passes through its spaces. Because it's also slower to drain, it has a tighter hold on plant nutrients. Clay soil is thus rich in plant food for better growth. Clay soil is cold and in the spring, takes time to warm since the water within also has to warm up.

4. Peaty

Peaty soil is dark brown or black in color, soft, easily compressed due to its high water content, and rich in organic matter. Their decay was so slow underwater that it led to the accumulation of organic area in a concentrated spot. Although peat soil tends to be heavily saturated with water, once drained, it turns into a good growing medium.

5. Saline Soil

The soil in extremely dry regions is usually brackish because of its high salt content, known as saline soil, it can cause damage to and stall plant growth, impede germination, and cause difficulties in irrigation. The salinity is due to the buildup of soluble salts in the rhizosphere–high salt contents prevent water uptake by plants, leading to drought stress.

6. Alluvial Soil

Materials deposited by rivers, winds, glaciers and sea waves are called alluvium and soils made up of alluvium are alluvial soils. In India alluvial soils are mainly found on the Indo-Ganga- Brahmaputra Plains, Coastal Plains and the broad river valleys of South India.

7. Red Soil

Red soils develop on granite and geneses rocks under low rainfall condition. The dissemination of red oxides of iron gives the characteristic red color of the soil.

8. Black Soil/Regur Soil

The regur is clayey, becomes very sticky when wet. Its special merit lies in its water holding capacity. These soils are very fertile and contain a high percentage of lime and a moderate amount of potash. The type of soil is especially suited to the cultivation of cotton and hence sometimes called 'black cotton soil.

9. Laterite Soil

Laterite is a kind of clayey rock or soil formed under high temperature and high rainfall. By further modification laterite is converted into red colored lateritic soils charged with iron nodules. Laterite and lateritic soils are found in South Maharashtra, the Western Ghats in Kerala and Karnataka, at places on the Eastern Ghat, in some parts of Assam, Tamil Nadu, Karnataka, and in western West Bengal (particularly in Birbhum district).

10. Desert/Arid Soil

The soils of Rajasthan, Haryana and the South Punjab are sandy. In the absence of sufficient wash by rain water soils have become saline and rather unfit for cultivation. In spite of that cultivation can be carried on with the help of modern irrigation. Wheat, bajra, groundnut, *etc.* can be grown in this soil.

11. Forest Soil

These are soil which are present in forest land. It is the mixture of soil and parts of plant. Humus content of this soil is high.

12. Mountain Soil

These are the soils which are found in mountains. Alluvium is found at the valley floor, brown soil, rich in organic matter, in an altitudinal zone lying between about 700-1800 m. Further up podzol soils, grey in color and acidic in reaction, are found associated with coniferous vegetation. In the Alpine forest belts the soils are thin and darker in color. This type of soil is suitable for the cultivation of potatoes, fruits, tea coffee and spices and wheat.

Soil Preparation

Soil preparation is one of the most important steps to having a successful vegetable garden. The best garden soil is fertile, well drained yet retains moisture, and gets enough air circulation which is needed for healthy roots. Find more information on what are your soil types.

Early spring is the time to start preparing your soil. Starting with soil testing will tell you how much nitrogen, phosphorus, and potassium you have in your soil and what the pH (a measure of hydrogen ion). From these results you can tell what soil amendments are needed for the best plant growth.

The steps of soil preparation are as follow:

1. Remove any grass.
2. Plow, spade or rototill the area.
3. Break down any large clumps and soil types
4. Apply manure or compost and soil amendments.
5. Conduct soil test find out if your soil needs any of the basic nutrients and add accordingly.
6. Add lime
7. Rake the bed: Pick out any large debris or small stones.

1.8 Irrigation Methods

Irrigation has been around for as long as humans have been cultivating plants, this is process of giving water to plants. Pouring water on fields is still a common irrigation method today — but other, more efficient and mechanized methods are also used. Irrigation is the method in which a controlled amount of water is supplied to plants at regular intervals for agriculture.

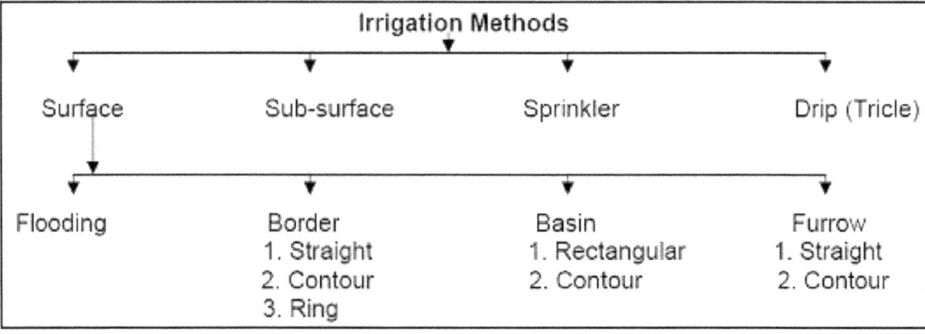

There are four principle methods of irrigation *viz.* surface, sub surface and aerial, overhead or sprinkler irrigation.

A. Surface irrigation

There are four variations under this method *viz.*

1. Flooding
2. Bed or border method (Saras and flat beds)
3. Basin method (ring and basin)
4. Furrow method (rides and furrows, broad ridges or raised beds)

(1) Flooding

It consist of opening a water channel in a plot or field so that water can flow freely in all directions and cover the surface of the land in a continuous sheet. It is the most inefficient method of irrigation as only about 20 per cent of the water is actually used by plants. It is suitable for uneven land where the cost of leveling is high and where a cheap and abundant supply of water is available. It is unsuitable for crops that are sensitive to water logging the method suitable where broadcast crops, particularly pastures, alfalfa, peas and small grains are produced.

Advantages

1. Can be used on shallow soils.
2. Can be employed where expense of leveling is great.
3. Installation and operation costs are low.
4. System is not damaged by livestock and does not interfere with use of farm implements.

Disadvantages

1. Excessive loss of water by run of and deep percolation.
2. Excessive soil erosion on step land.
3. Fertilizer and Farm Yard Manure are eroded from the soil.

(2) Bed or Border Method (Sara and Flat beds or check basin)

In this method the field is leveled and divided into small beds surrounded by bunds of 15 to 30 cm high. Small irrigation channels are provided between two adjacent rows of beds. The length of the bed varies from 30 meters for loamy soils to 90 meters for clayey soils. The width is so adjusted as to permit the water to flow evenly and wet the land uniformly. For high value crops, the beds may be still smaller especially where water is costly and not very abundant. This method is adaptable to most soil textures except sandy soils and is suitable for high value crops. It requires leveled land. It is more efficient in the use of water and ensures its uniform application. It is suitable for crops plant in lines or sown by broadcast. This may also be called a sort of sara method followed locally in Maharashtra but the saras to be formed in this method are much longer than broader.

Advantages

1. Fairly large supply of water is needed.
2. Land must be leveled
3. Suited only to soils that do not readily disperse.
4. Drainage must be provided

(3) Basin Irrigation

This method is suitable for orchids and other high value crops where the size of the plot to be irrigated is very small. The basin may be square, rectangular or circular shape. A variation in this method *viz.* ring and basin is commonly used for irrigating fruit trees. A small bund of 15 to 22 cm high is formed around the stump of the tree at a distance of about 30 to 60 cm to keep soil dry. The height of the outer bund varies depending upon the depth of water proposed to retain. Basin irrigation also requires leveled land and not suitable for all types of soil. It is also efficient in the use of water but its initial cost is high.

Advantages

1. Varying supply of water
2. No water loss by runoff
3. Rapid irrigation possible
4. No loss of fertilizers and organic manures
5. Satisfactory results

Disadvantages

1. If land is not leveled initial cost may be high
2. Suitable mainly for orchids, rice, jute *etc.*
3. Except rice, not suitable for soils that disperse easily and readily from a crust.

(4) Furrow Method (Rides and Furrow, Broad Ridges, Counter Furrow *etc.*)

Water is allowed to flow in furrow opened in crop rows. It is suitable for sloppy lands where the furrows are made along contours. The length of furrow is determined mostly by soil permeability. It varies from 3 to 6 meters. In sandy and clay loams, the length is shorter than in clay and clay loams. Water does not come in contact with the plant stems. There is a great economy in use of water. Some times, even in furrow irrigation the field is divided into beds having alternate rides and furrows. On slopes of 1 to 3 per cent, furrow irrigation with straight furrows is quite successful. But on steeper slopes contour furrows, not only check erosion but ensure uniform water penetration.

Advantages

1. High water efficiency
2. Can be used in any row crop
3. Relatively easy in stall
4. Not expensive to maintain
5. Adapted to most soils.

Disadvantages

1. Requirement of skilled labour
2. Hazard to operation of machinery
3. Drainage must be provided.

B. Subsurface Method

Subsurface irrigation or sub-irrigation may be natural or artificial. Natural sub surface irrigation is possible where an impervious layer exists below the root zone. Water is allowed in to series of ditches dug up to the impervious layer, which then moves laterally and wets root zone. It is very efficient in the use of water as evaporation is cut off almost completely. The plant roots do not suffer from logging, there is no loss of agricultural land in laying out irrigation system and implements can be worked out freely.

C. Drip or Trickle Irrigation

It involves slow application of water to the root zone. The drip irrigation system consist of:

1. Head
2. Main line and sub line
3. Lateral lines
4. Drip nozzles.

The head consists of a pump to lift water and produce the desired pressure (about 2.5 atmosphere) and to distribute water through nozzles. Mains and sub mains are normally of flexible material such as black PVC pipes. Laterals or drip lines are small diameter flexible lines (usually 1 to 1.25 cm diameter black PVC tubes) taking off from the mains or sub mains. Laterals are normally laid parallel to each other. Lateral lines can be up to about 50 meters long and are usually 1.2 cm diameter black plastic tubing. There is usually one lateral line for each crop row. By laying the main line along the center line of the field, it is possible to irrigate either side of the field alternately by shifting the laterals. A pressure drop of 10 per cent is permitted between the ends of lateral. Drip nozzles are also known as emitters or values and are fixed at regular intervals in the laterals. These PVC values allow water to flow at the extremely slow rates, ranging from 2 to 11 liters per hour and they are of different shapes and design. The spacing between laterals is controlled by the row-to-row spacing of the crop to be irrigated. Drip laterals laid on soil surface are buried underground at the depth of 5 to 10 cm.

Advantages

1. The losses by drip irrigation and evaporation are minimized
2. Precise amount of water is applied to replenish the depleted soil moisture at frequent intervals for optimum plant growth.
3. The system enables the application of water fertilizers at an optimum rate to the plant root system.
4. The amount of water supplied to the soil is almost equal to the daily consumptive use, thus maintaining a low moisture tension in soil.

Disadvantages

The initial cost of the drip irrigation for large-scale irrigation is its main limitation. The cost of the unit per hectare depends mainly on the spacing of the crop. For widely spaced crops like fruit trees, the system may be even more economical than sprinkler.

D. Sprinkler or Overhead Irrigation

This method consists of application of water to soil in the form of spray, somewhat as rain. It is particularly useful for sandy soils because they absorb water too fast. Soils that are too shallow, too steep or rolling can be irrigated efficiently with sprinklers.

This method is suitable for areas having uneven topography and where erosion hazards are great.

In sprinkler irrigation, water is conveyed under pressure through pipes to the area to be irrigated where it is passed out through or sprinklers the system comprises four main parts:

1. Power generator
2. Pump
3. Pipeline
4. Sprinkler

The power generator may be electrical or mechanical. A centrifugal pump may be used for suction lift up to 37 to 50 cm. A piston type pump is preferable where water is very deep. The pipe consists of two sections, the main line and the laterals.The main line may be permanently buried underground or may be laid above ground, if it is to be used on a number of fields. The main pipes are usually made of steel or iron.

The laterals are lightweight aluminum pipes and are usually portable. The sprinkler nozzles may be single or double, revolving or stationery and mounted or riser pipes attached to riser. Each sprinkler head applies water to circular area whose diameter depends up on the size of water, which varies from ¼ to ¾ inch per hour is determined by selecting the proper combination of nozzles.

Advantages

1. It ensures uniform distribution of water
2. It is adaptable to most kinds of soil.
3. It offers no hindrance to the use of farm implements
4. Fertilizers material may be evenly applied through sprinklers. This is done by drawing liquid fertilizer solution slowly in to the pipes on the suction side of the pump so that the time of application varies from 10 to 30 minutes.
5. Water losses are reduced to a minimum extent
6. More land can be irrigated
7. Costly land leveling operations are not necessary and
8. The amount of water can be controlled to meet the needs of young seedling or mature crops.

Disadvantages

1. The initial cost is rather very high.
2. Any cost of power to provide pressure must be added to the irrigation charges.
3. Wind interferes with the distribution pattern, reducing spread or increasing application rate near lateral pipe.
4. There is often trouble from clogged nozzle or the failure of sprinklers to revolve.
5. The cost of operations and maintaince is very high. Labour requirement for moving a pipe and related work approximately nearly one hour per irrigation.
6. It requires a dependable constant supply of water free slit and suspended matter.
7. It is suitable for high value crops

1.9. Hydroponics/Soilless Culture

Hydroponics is a method of growing crops without soil. Hydroponics is very simple — in many ways, it's simpler than growing plants in soil. Hydroponics is a subset of hydroculture, the method of growing plants without soil, using mineral nutrient solutions in a water solvent. Terrestrial plants may be grown with only their roots exposed to the mineral solution, or the roots may be supported by an inert medium, such as perlite or gravel. The nutrients in hydroponics can come from an array of different sources; these can include but are not limited to waste from fish waste, duck manure, or normal nutrients.

Types of Hydroponic Systems

a. Nutrient Film Technique (NFT)

There are several variations of N.F.T. used around the world and it is a very popular method of growing hydroponically. Nutrient Film Technique uses a constant flow of your Growth Technology nutrient solution (therefore no timer is required). The solution is pumped from a reservoir into the growing tray. The growing tray requires no growing medium. The roots draw up the nutrients from the flowing solution. The downward flow pours back into the reservoir to be recycled again. Pump and electric maintenance is essential to avoid system failures, where roots can dry out rapidly when the flow stops.

b. Wicks System

The Wick system is described as a passive system, by which there are no moving parts. From the bottom reservoir, Growth Technology nutrient solution is drawn up through a number of wicks into the growing medium.

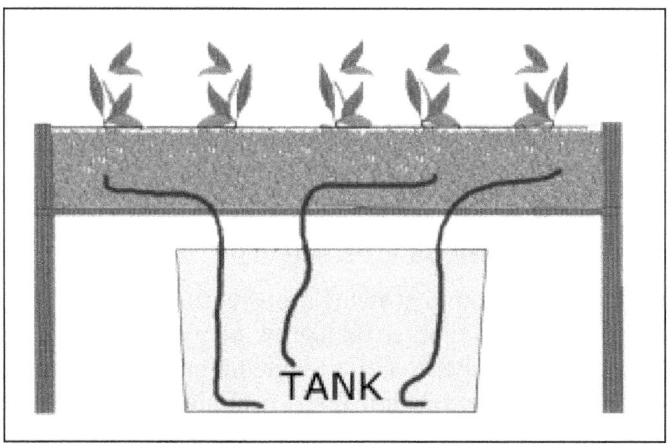

c. Water Culture

This system is an active system with moving parts. As active hydroponic systems go, water culture is the simplest. The roots of the plant are totally immersed in the water which contains the specific Growth Technology nutrient solutions. An air pump with help oxygenate the water and allow the roots to breathe.

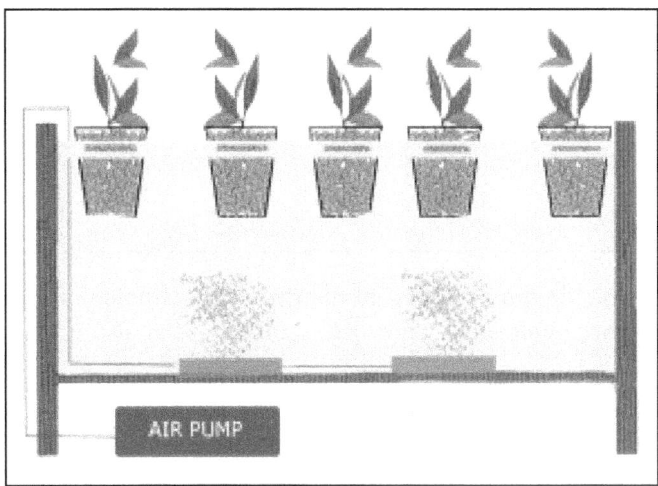

d. Ebb and Flow System (Flood and Drain)

This hydroponic system works by temporarily flooding the grow tray. The nutrient solution from a reservoir surrounds the roots before draining back. This action is usually automated with a water pump on a timer.

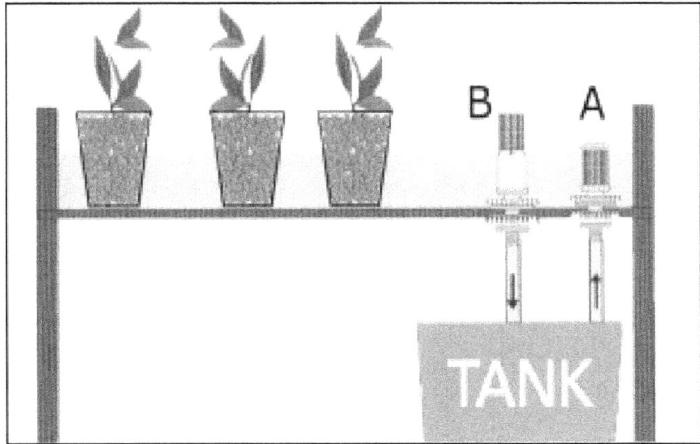

e. Drip System (Recovery or non-recovery)

Dip systems are a widely used hydroponic method. A timer will control a

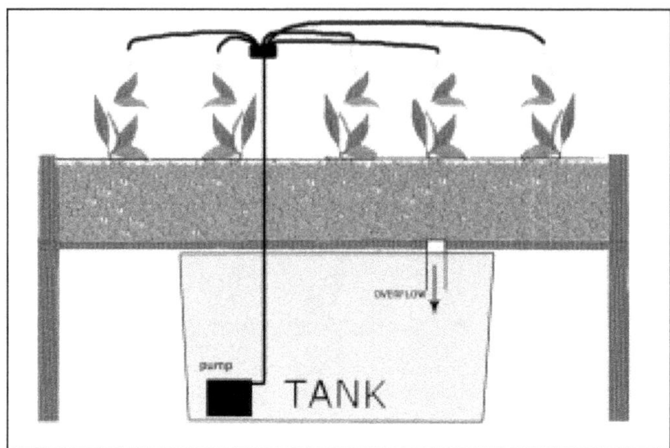

water pump, which pumps water and the Growth Technology nutrient solutions through a network of elevated water jets, it supplies water to the growing plants like that of a drip.

Chapter 2

PROPAGATION METHODS

Plant propagation is the process of creating new plants. There are two types of propagation, they are sexual and asexual. Sexual reproduction is the union of the pollen and egg, drawing from the genes of two parents to create a new, third individual. Sexual propagation involves the floral parts of a plant. Asexual propagation involves taking a part of one parent plant and causing it to regenerate itself into a new plant. The resulting new plant is genetically identical its parent. Asexual propagation involves the vegetative parts of a plant: stems, roots or leaves. Propagation of plant is defined as production of new individuals from a selected plant having all the characters of the original one.

Importance of Plant Propagation

1. Multiply the different species in large number.
2. Protect the plant species which are endangered.
3. Improve the characteristics and quality of the plants.
4. Produce quality and healthy plants on commercial base.

2.1. Plant Propagation Types

a. Sexual Propagation

As the name suggests, sexual propagation involves contribution of both female and male sexes for creation of new plants. It is a natural process in which a parent species create offspring that are genetically different from them.

Advantages of Sexual Propagation

- ☆ Simplest, easiest and the most economical process among various types of plant propagation.
- ☆ Some plants, trees, vegetables or fruits species can propagate only through sexual propagation. *E.g.* – marigold, papaya, tomato.
- ☆ Sexual propagation leads to better crop species that are stronger, disease-resistant and have longer life-span.
- ☆ Viral transmission can be prevented in this type of propagation.
- ☆ Sexual propagation is responsible for production of large number of crops and that too with different varieties.
- ☆ It is the only propagation process in which resultant offspring have genetic variation and exhibit diversity of characters from parent crops. This genetic variation is responsible for continuous evolution that keeps on producing better and better offspring.
- ☆ Easy storage and transportation of seeds.
- ☆ This is very simple and easy method of propagation.

b. Asexual Propagation

Also known as vegetative propagation, this process involves production of species through vegetative parts of the plants such as roots, leaves, stems, bulbs, tubers *etc.* In this process, no exchange of genetic information takes place as the offspring is formed through material of a single parent. Thus the resultant plants formed are identical to the parent plant (also known as clones).

Different Asexual organs for propagation are:

a. Stolons and Runners

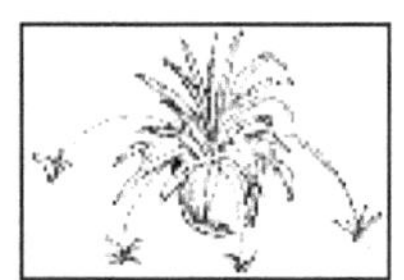

A stolon is a horizontal, often fleshy stem that can root, then produce new shoots where it touches the medium. A runner is a slender stem that originates in a leaf axil and grows along the ground or downward from a hanging basket, producing a new plant at its tip. Plants that produce stolons or runners are propagated by severing the new plants from their parent stems. Plantlets at the tips of runners may be rooted while still attached to the parent, or detached and placed in a rooting medium.

Examples: Strawberry, Spider plant.

b. Offsets

Plants with a rosetted stem often reproduce by forming new shoots at their base or in leaf axils. Sever the new shoots from the parent plant after they have developed their own root system. Unrooted offsets of some species may be removed and placed in a rooting medium. Some of these must be cut off, while others may be simply lifted off the parent stem.

Examples: Date palm, Haworthia, Bromeliads, Many cacti.

c. Separation

Separation is a term applied to a form of propagation by which plants that produce bulbs or corms for multiplication.

d. Bulbs

These are the bulb like propagules. Separate these bulb clumps every 3 to 5 years for largest blooms and to increase bulb population. Dig up the clump after the leaves have withered. Gently pull the bulbs apart and replant them immediately so their roots can begin to develop. Small, new bulbs may not flower for 2 or 3 years, but large ones should bloom the first year.

Examples: Tulip, Marcissus.

e. Corms

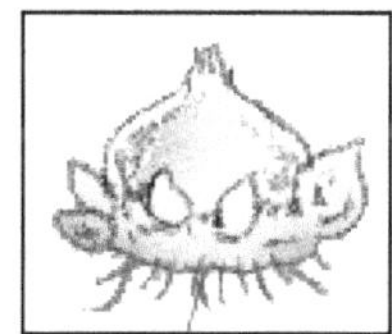

These are vegetative propagules and tiny cormels form around the large corm. After the leaves wither, dig up the corms and allow them to dry in indirect light for 2 or 3 weeks. Remove the cormels, then gently separate the new corm from the old corm. Dust all new corms with a fungicide and store in a cool place until planting time.

Examples: Crocus, Gladiolus.

f. Crowns

Plants with more than one rooted crown may be divided and the crowns planted separately. Divisions of some outdoor plants should be dusted with a fungicide before they are replanted.

Examples: Snake plant, Iris, Prayer plant, Day lilies.

Propagation by cuttings, division, layering and grafting/budding are various methods of asexual propagation process.

Advantages of Asexual Propagation

- ☆ As resultant species formed through asexual process are genetically identical, useful traits can be preserved among them.
- ☆ Asexual propagation allows propagation of crops that do not possess seeds or those which are not possible to grow from seeds. For *e.g.* Jasmine, sugarcane, potato, banana, rose *etc.*
- ☆ Plants grown through vegetative propagation bear fruits early.

- ☆ In this type, only a single parent is required and thus it eliminates the need for propagation mechanisms such as pollination, cross pollination *etc.*
- ☆ The process is faster than sexual propagation. This helps in rapid generation of crops which in turn balances the loss.
- ☆ Injured plants can be recovered or repaired through techniques involved in asexual propagation.

2.2. Cuttings

a. Stem Cutting

Propagation by stem cuttings is the most popular plant propagation method for woody shrubs and ornamental plants. This is also a good technique for houseplants. For stem cutting remove the bottom set of leaves of the stem and cut it into proper and dip the end you just cut into rooting gel. This will help seal the cut plant tissue and promote new root growth. Then place the cutting into a small pot with moist vermiculite, perlite or other soilless potting mix. Once the cuttings have developed roots – this can take a few days or a few months – replant them in another container with moist, but not wet, potting soil.

Different Stem cuttings are:

1. Tip Cuttings

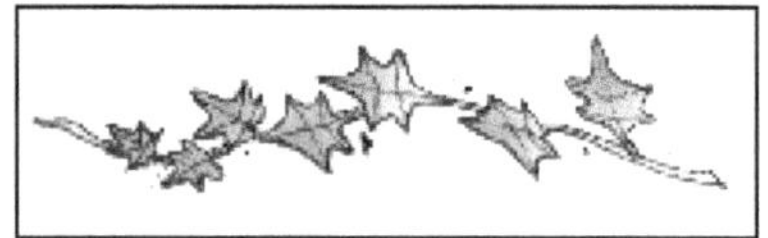

Detach a 2- to 6-inch piece of stem, including the terminal bud. Make the cut just below a node. Remove lower leaves that would touch or be below the medium. Dip the stem in rooting hormone if desired. Gently tap the end of the cutting to remove excess hormone. Make a hole in the medium with a pencil or pot label, and insert the cutting deeply enough into the media to support itself.

2. Medial Cuttings (also stem-section cuttings)

Make the first cut just above a node, and the second cut just below a node 2 to 6 inches down the stem. Prepare and insert the cutting as you would a tip cutting. Be sure to position right side up. Buds are always above leaves. Make sure the cutting is inserted base down.

3. Cane Cuttings

Cut cane-like stems into sections containing one or two eyes, or nodes. Dust ends with fungicide or activated charcoal. Allow to dry several hours. Lay horizontally with about half of the cutting below the media surface, eye facing upward. Cane cuttings are usually potted when roots and new shoots appear, but new shoots from dracaena and croton are often cut off and re rooted in sand.

4. Single Eye

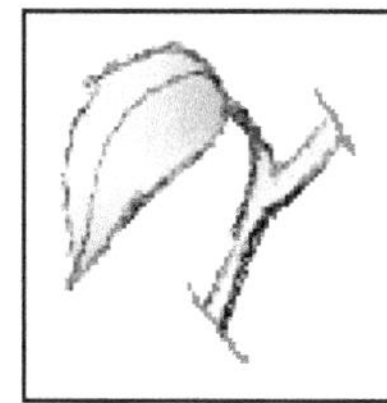

The eye refers to the bud which emerges at the axil of the leaf at each node. This is used for plants with alternate leaves when space or stock material are limited. Cut the stem about ½ inch above and ½ inch below a node. Place the cutting horizontally or vertically in the medium with the node just touching the surface.

5. Double Eye

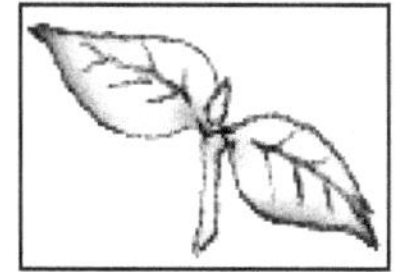

This is used for plants with opposite leaves when space or stock material is limited. Cut the stem about ½ inch above and ½ inch below the same node. Insert the cutting vertically in the medium with the node just touching the surface.

6. Heel Cutting

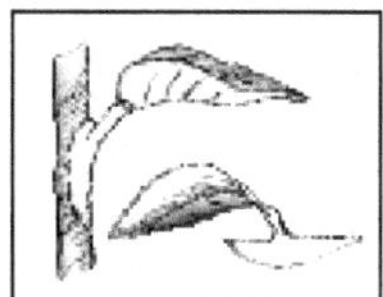

This method uses stock material with woody stems efficiently. Make a shield-shaped cut about halfway through the wood around a leaf and axial bud. Insert the shield horizontally into the medium so that it is completely covered. Remove any leaf blade but keep a portion of the petiole intact for ease in handling this small cutting.

b. Leaf Cuttings

Several herbaceous or woody plants, including many indoor houseplants, can be propagated from leaf cuttings. With this method, a leaf and its stem (petiole) or sometimes just a piece of the leaf are used to create an entirely new plant. The directions for propagation by leaf cuttings are basically the same as for softwood or hardwood stem cuttings and can be performed any time of year.

Select a healthy, full grown leaf from a vigorously growing plant and remove it along with about 1½ inches of its stem. Dip the cut portion in rooting hormone and plant the entire stem (up to the bottom of the leaf) at an angle in a moist soilless rooting medium. After planting water thoroughly to settle the potting mix around the plant.

Types of Leaf Cuttings are:

1. Whole Leaf with Petiole

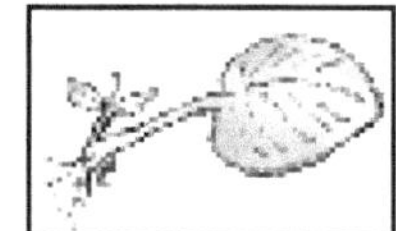

Detach the leaf and up to 1½ inches of petiole. Insert the lower end of the petiole into the medium. One or more new plants will form at the base of the petiole. The leaf may be severed from the new plants when they have their own roots, and the petiole can be reused. (Example: African violet).

2. Whole Leaf without Petiole

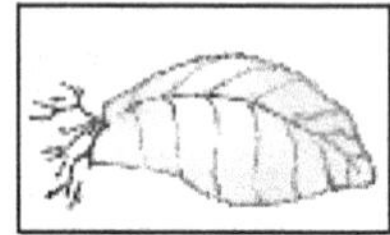

This is used for plants with sessile leaves (no stalk or petiole). Insert the cutting vertically into the medium. A new plant will form from the axillary bud. The leaf may be removed when the new plant has its own roots. (Example: Donkey's tail).

3. Split Vein

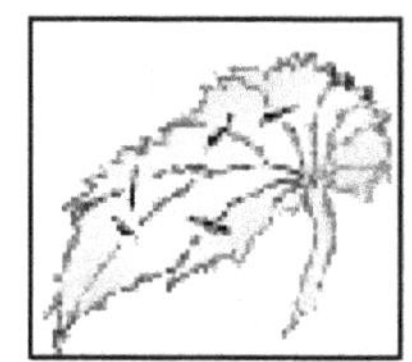

Detach a leaf from the stock plant. Slit its veins on the lower leaf surface. Lay the cutting, lower side down, on the medium. New plants will form at each cut. If the leaf tends to curl up, hold it in place by covering the margins with the rooting medium. (Example: Rex begonia).

4. Leaf Sections

This method is frequently used with snake plant and fibrous rooted begonias. Cut begonia leaves into wedges with at least one vein. Lay leaves flat on the medium. A new plant will arise at the vein. Cut snake plant leaves into 2-inch sections. Consistently make the lower cut slanted and the upper cut straight so you can tell which is the top. Insert the cutting vertically. Roots will form fairly soon, and eventually a new plant will appear at the base of the cutting. These and other succulent cuttings will rot if kept too moist.

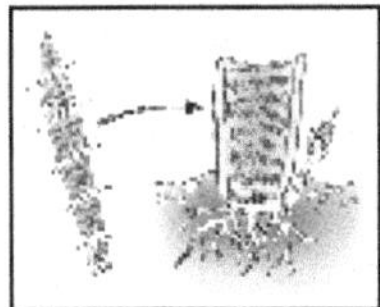

c. Root Cutting

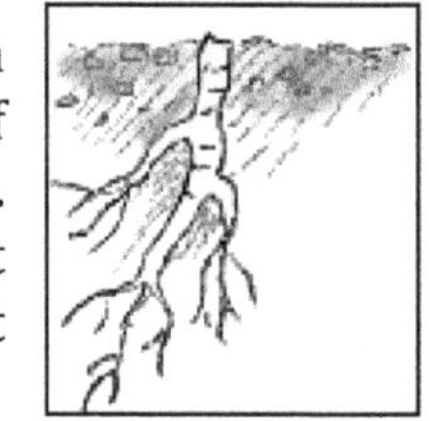

Take 1 to 4-inch long cuttings from younger root growth that is about ¼ to ½ inch thick. Cut straight through the end of the root nearest to the stem and cut the other end at an angle. This way you will remember which end is the top (the straight cut) and which is the bottom (the diagonal cut). Roots will not grow if you plant them upside down.

Root cuttings are usually taken from 2- to 3-year-old plants during their dormant season when they have a large carbohydrate supply. Root cuttings of some species produce new shoots, which then form their own root systems, while root cuttings of other plants develop root systems before producing new shoots

Types of Root Cutting are:

a. Plants with Large Roots

Make a straight top cut. Make a slanted cut 2 to 6 inches below the first cut. Store about 3 weeks in moist sawdust, peat moss, or sand at 40°F. Remove from storage. Insert the cutting vertically with the top approximately level with the surface of the rooting medium. This method is often used outdoors. (Example: Horse radish).

Plants with Small Roots

Take 1- to 2-inch sections of roots. Insert the cuttings horizontally about ½ inch below the medium surface. This method is usually used indoors or in a hotbed. (Example: Bleeding heart).

2.3 Layering

Layering is a way to grow new plants from existing plants without having to take any cuttings. In a nutshell, bury part of a stem or branch in the soil and new roots and shoots will form. This method is often more successful than propagating from cuttings, because the new plant can get water and food from the stock plant. Once the new plant is established, it can be moved to another spot in the garden.

a. Tip Layering

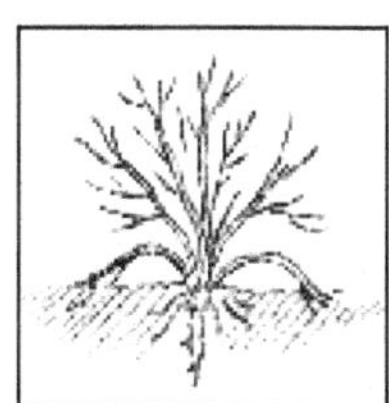

Dig a hole 3 to 4 inches deep. Insert the shoot tip and cover it with soil. The tip grows downward first, then bends sharply and grows upward. Roots form at the bend, and the recurved tip becomes a new plant. Remove the tip layer and plant it in the early spring or late fall.

Examples: Purple and black raspberries, trailing blackberries.

b. Simple Layering

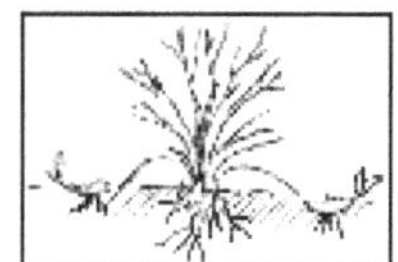

Bend the stem to the ground. Cover part of it with soil, leaving the last 6 to 12 inches exposed. Bend the tip into a vertical position and stake in place. The sharp bend will often induce rooting, but wounding the lower side of the branch or loosening the bark by twisting the stem may help.

Examples: Forsythia, Honeysuckle.

c. Compound Layering or Serpentine Layering

This method works for plants with flexible stems. Bend the stem to the rooting medium as for simple layering, but alternately cover and expose stem sections. Wound the lower side of the stem sections to be covered. Examples: Heart-leaf philodendron, Pothos.

d. Mound (Stool) Layering

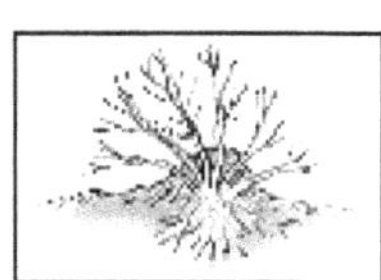

Cut the plant back to one inch above the ground in the dormant season. Mound soil over the emerging shoots in the spring to enhance their rooting.

Examples: Gooseberries, Apple rootstocks.

e. French Layering

It involves cutting back the parent plant hard in spring to produce lots of new stems near ground level. The following spring, these new shoots are pegged down on to the soil like the spokes of a wheel radiating out from the base of the plant. As side shoots grow upwards from these stems, soil is mounded over them to encourage rooting. By the autumn or the following spring, these rooted sections can be separated and planted out independently.

Examples: *Cotinus*, *Cornus* and *Prunus tenella*.

f. Air Layering

Air layering can be used on many larger houseplants, as well as woody ornamental plants, such as holly, rhododendron and lilac. Using an upper branch or stem, select a site just below a leaf node and remove the leaves and twigs both below and above that point for 3-4 inches. Scrape away a small area of bark, or make a cut about 1½ inches long and $^1/_3$ of the way through the stem. Apply a rooting mixture, to the exposed area to promote root production. Use roughly a handful of moist sphagnum moss to surround the wound and wrap the moss with black plastic. Seal the plastic on all sides with tape or twisty ties, making sure that the moss does not extend beyond the cover.

Once the roots are well formed (usually 1-3 months for houseplants; 1-2 seasons for outdoor plants) cut the stem just below the bag and pot the new plant as you would any seedling. After a couple of months the young plant should be hardy enough to transplant outside. Click on this link to learn more about air layering for difficult to root plants.

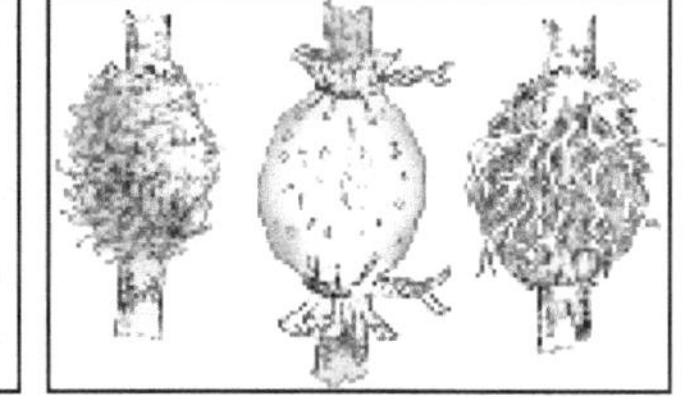

Air layering is used to propagate some indoor plants with thick stems, or to rejuvenate them when they become leggy. Slit the stem just below a node. Pry the slit open with a toothpick. Surround the wound with wet sphagnum moss. Wrap plastic or foil around the sphagnum moss and tie in place. When roots pervade the moss, cut the plant off below the root ball.

2.4 Grafting

"Stock" is a rooted plant upon which a branch of a desired variety of the plant is grafted. The branch, which is being grafted, is called as "scion". Grafting is done on a stock plant, which has a very strong root system. Mango, Chikoo and Golden Champa Jackfruit and Jamun plants are being successfully being grafted.

Types of Grafting

a. Cleft Grafting

One of the simplest and most popular forms of grafting, cleft grafting), is a method for top working both flowering and fruiting trees (apples, cherries, pears, and peaches) in order to change varieties. Cleft grafting may be performed on main stems or on lateral or scaffold branches. The rootstock used for cleft grafting should range from 1 to 4 inches in diameter and should be straight grained. The scion should be about ¼ inch in diameter, straight, and long enough to have at least three buds. Scions that are between 6 and 8 inches long are usually the easiest to use. Insert a scion on each end of the cleft, with the wider side of the wedge facing outward. The cambium of each scion should contact the cambium of the rootstock. Pressure from the rootstock will hold the scions in place. Thoroughly seal all cut surfaces with grafting wax or grafting paint to keep out water and prevent drying. If both scions in the cleft "take," one will usually grow more rapidly than the other.

b. Bark Grafting

Bark grafting is used primarily to top work flowering and fruiting trees. In contrast to cleft grafting, this technique can be applied to rootstock of larger diameter (4 to 12 inches) and is done during early spring when the bark slips easily from the wood but before major sap flow. The rootstock is severed with a sharp saw, leaving a clean cut as with cleft grafting. Start at the cut surface of the rootstock and make a vertical slit through the bark where each scion can be inserted (2 inches long and spaced 1 inch apart). Since multiple scions are usually inserted around the cut surface of the rootstock, prepare several scions for each graft. Cut the base of each scion to a 1½ to 2-inch tapered wedge on one side only. Loosen the bark slightly and insert the scion so that the wedge-shaped tapered surface of the scion is against the exposed wood under the flap of bark. Push the scion firmly down into place behind the flap of bark, replace the bark flap, and nail the scion in place by driving one or two wire brads through the bark and scion into the rootstock. Insert a scion every 3 to 4 inches around the cut perimeter of the rootstock. Seal all exposed surfaces with grafting wax or grafting paint. Once the scions have begun to grow, leave only the most vigorous one on each stub; prune out all the others. Bark grafts tend to form weak unions and therefore usually require staking or support during the first few years.

c. Side-Veneer Grafting

At one time the side-veneer graft was a popular technique for grafting varieties of camellias and rhododendrons that are difficult to root. Currently, it is the most popular way to graft conifers, especially those having a compact or dwarf form. Side-veneer grafting is usually done on potted rootstock. Rootstock is grown in pots the season before grafting, allowed to go dormant, and then stored as with other container nursery stock. After exposure to cold weather for at least six weeks, the rootstock is brought into a cool greenhouse for a few days before grafting takes place to encourage renewed root growth. The plant should not be watered at this time. Make a shallow downward cut about ¾ inch to 1 inch long at the base of the stem on the potted rootstock to expose a flap of bark with some wood still attached. Make an inward cut at the base so that the flap of bark and wood can be removed from the rootstock. Choose a scion with a diameter the same as or slightly smaller than the rootstock. Make a sloping cut ¾ inch to 1 inch long at the base of the scion. Insert the cut surface of the scion against the cut surface of the rootstock. Be certain that the cambia contact each other. Hold the scion in place using a rubber grafting strip, tape, or grafting twine. Seal the entire graft area with warm grafting wax or grafting paint. Remove the rubber or twine shortly after the union has healed. Never allow the binding material to girdle the stem.

d. Splice Grafting

Splice grafting is used to join a scion onto the stem of a rootstock or onto an intact root piece. This simple method is usually applied to herbaceous materials that callus or "knit" easily, or it is used on plants with a stem diameter of ½ inch or less. In splice grafting, both the stock and scion must be of the same diameter. Cut off the rootstock using a diagonal cut ¾ inch to 1 inch long. Make the same type of cut at the base of the scion. Fit the scion to the stock. Wrap this junction securely with a rubber grafting strip or twine. Seal the junction with grafting wax or grafting paint. Water rootstock sparingly until the graft knits. Over watering may cause sap to "drown" the scion. Be sure to remove the twine or strip as soon as the graft has healed.

e. Whip and Tongue Grafting

The whip and tongue technique is most commonly used to graft nursery crops or woody ornamentals. Both the rootstock and scion should be of equal size and preferably no more than ½ inch in diameter. The technique is similar to splice grafting except that the whip on the rootstock holds the tongue of the scion in place (and vice versa). This leaves both hands free to wrap the joint. For the whip and tongue graft, make similar cuts on both the stock and scion. These cuts should be made with a single draw of the knife and should have a smooth surface so that the two can develop a good graft union. Up to this point, rootstock and scion are cut the same as for a splice graft. Cut off the stock using a diagonal cut. The cut should be four to five times longer than the diameter of the stock to be grafted. Make the same kind of cut at the base of the scion. Next, place the blade of the knife across the cut end of the stock, halfway between the bark and pith (on the upper part of the cut surface). Use a single knife stroke to draw the blade down at an angle through the wood and pith. Stop at the base of the initial diagonal cut. This second cut must not follow the grain of the wood but should run parallel to the first cut. Prepare the scion in the same way. Fit the scion into the rootstock so that they interlock whip and tongue. Be certain that the cambia are aligned. Wrap the junction with a grafting strip or twine, and seal it with grafting wax or grafting paint. Never allow the binding material to girdle the stem.

f. Saddle Grafting

Saddle grafting is a relatively easy technique to learn and once mastered can be performed quite rapidly. The stock may be either field-grown or potted. Both rootstock and scion should be the same diameter. For best results, use saddle grafting on dormant stock in mid- to late winter. Stock should not be more than 1 inch in diameter. Using two opposing upward strokes of the grafting knife, sever the top from the rootstock. The resulting cut should resemble an inverted V, with the surface of the cuts ranging from ½ inch to 1 inch long. Now reverse the technique to prepare the base of the scion. These cuts on the rootstock and scion must be the same length and have the same slope so that a maximum amount of cambial tissue will make contact when the two halves are joined. Place the V-notched scion onto the saddle of the rootstock. If rootstock and scion are the same diameter, cambial alignment is easier; otherwise adjust as needed. Wrap the graft with a grafting twine, tape, or strip, then seal it with grafting wax or grafting paint.

g. Bridge Grafting

Bridge grafting is used to "bridge" a diseased or damaged area of a plant, usually at or near the base of the trunk. Such damage commonly results from contact with grading or lawn maintenance equipment, or it may be caused by rodents, cold temperatures, or disease organisms. The bridge graft provides support as well as a pipeline that allows water and nutrients to move across the damaged area. Bridge grafts are usually done in early spring just before active plant growth begins. They may be performed any time the bark on the injured plant "slips." Select scions that are straight and about twice as long as the damaged area to be bridged. Make a 1½ to 2-inch-long tapered cut on the same plane at each end of the scion. Remove any damaged tissue so the graft is on healthy stems. Cut a flap in the bark on the rootstock the same width as the scion and below the injury to be repaired. Gently fold the flap away from the stock, being careful not to tear the bark flap. First, insert and secure the scion below the injury; push the scion under the flap with the cut portion of the scion against the wood of the injured stem or trunk. Then go back and insert and secure the scion above the injury following these same steps. Push the scion firmly into place. Pull the flap over the scion and tack it into place as described for bark grafting. When grafting with young stems that may waver in the wind, insert the scions so that they bow outward slightly. Bridge grafts should be spaced about 3 to 4 inches apart across the damaged area. Secure all graft areas with warm grafting wax or grafting paint. During and after the healing period, remove any buds or shoots that develop on the scions.

h. Inarch Grafting

Inarching, like bridge grafting, is used to bypass or support a damaged or weakened area of a plant stem. Unlike bridge grafting, the scion can be an existing shoot, sucker, or waterspout that is already growing below and extending above the injury. The scion may also be a shoot of the same species as the injured plant growing on its own root system next to the main trunk of the damaged tree. With the inarching technique, the tip of the scion is grafted in above the injury using the same method as for bark or bridge grafting.

2.5 Budding

Budding is a form of asexual reproduction in which a new organism develops from an outgrowth or bud due to cell division at one particular site. The new organism remains attached as it grows, separating from the parent organism only when it is mature, leaving behind scar tissue. Since the reproduction is asexual, the newly created organism is a clone and is genetically identical to the parent organism. T-budding, chip budding and patch budding are grafting techniques used for top-working trees and producing new plants. Use budding techniques for plants that cannot be propagated by seed or that have branches measuring one-quarter inch to one inch in diameter. A successful union through budding plants is only possible when using two compatible and common species.

a. T-budding or Shield Budding

T- budding or shield budding is a special grafting technique in which the scion piece is reduced to a single bud. As with other techniques of asexual propagation, the resulting plants are clones (genetically identical plants reproduced from one individual entirely by vegetative means). The plant being propagated (represented by the bud) is referred to as the scion, while the plant being grafted onto is referred to as the rootstock, or simply stock. A small branch with several buds suitable for T budding on it is often called a bud stick.

Rootstock can be grown in the field where it will be budded, or dormant liners can be transplanted into the field and then allowed to grow under moderate fertility until they reach the desired 3/16- to 7/16-inch caliper. Since budding is generally done less than 4 inches above the soil surface, leaves and side branches must be removed from this portion of the rootstock to create a clean, smooth working area. To avoid quickly dulling the knife, remove any soil from the rootstock where the cut will be made just before actual budding takes place. The stem can be cleaned by brushing or rubbing it gently by hand or with a piece of soft cloth.

Collect scion or budwood early in the day while temperatures are cool and the plants are still fully turgid. The best vegetative buds usually come from the inside canopy of the tree on the current season's growth. Mature buds are most desirable; discard terminal and younger buds because they are often not mature. To keep budwood from drying out, getting hot, or freezing (depending on the season), place it into plastic bags or wrap it in moist burlap as it is collected. Then

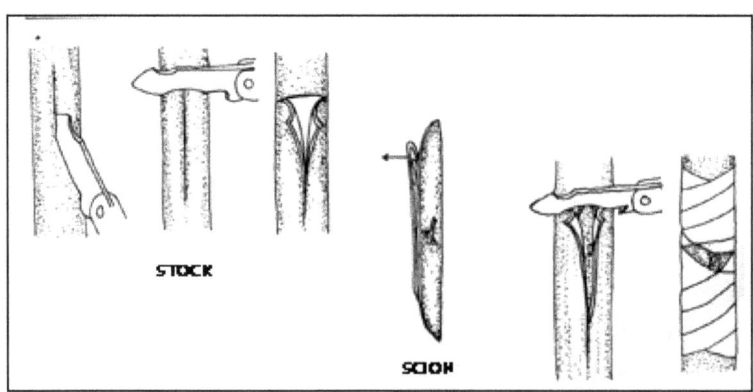

move to a shaded or sheltered area to prepare the buds. Place budwood of only one variety in each labeled bag.

Budsticks are usually prepared in a cool, shaded area. Remove the leaves but keep the petioles (leaf stem) intact to serve as handles when inserting a bud into the rootstock. Then cut the sticks to a convenient length, leaving three to six buds per stick. Budsticks that will not be used immediately should be bundled, labeled, and stored in moisture-retaining containers such as plastic bags or waxed cardboard boxes and kept cool (32° to 45°F). The longer budwood is stored, the less likely it is to "take." Generally, budwood stored for more than a few days should be discarded. When budwood is taken to the field, equal precautions against drying should be taken. Storing budwood in a picnic cooler with ice will help keep it cool and moist. Individual bundles of scions carried by budders are often wrapped in moist burlap or kept in dark (not clear) plastic.

b. Chip Budding

Chip budding is a technique that may be used whenever mature buds are available. Because the bark does not have to "slip," the chip-budding season is longer than the T-budding season. Species whose bark does not slip easily without tearing may be propagated more successfully by chip budding than by T-budding.

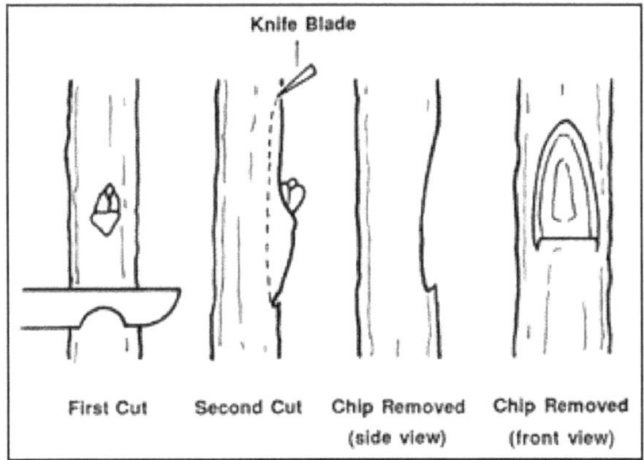

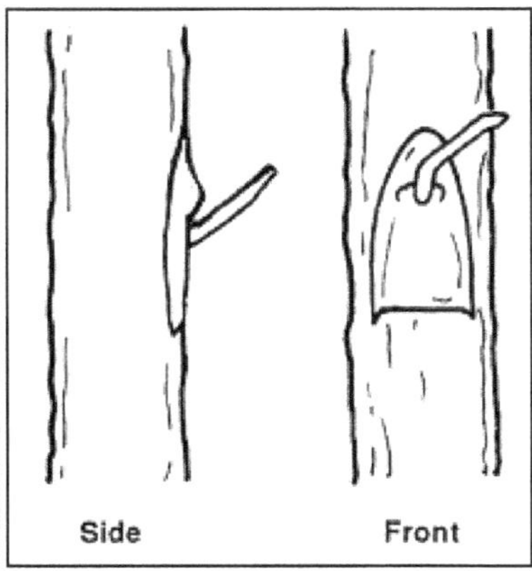

Although all the basics in handling budwood and stock are the same for chip budding and T budding, the cuts made in chip budding differ radically. The first cut on both stock and scion is made at a 45° to 60° downward angle to a depth of about $^{1}/_{8}$ inch. After making this cut on a smooth part of the rootstock, start the second cut about ¾ inch higher and draw the knife down to meet the first cut. (The exact spacing between the cuts varies with species and the size of the buds.) Then remove the chip. Cuts on both the scion (to remove the bud) and the rootstock (to insert the bud) should be exactly the same. Although the exact location is not essential, the bud is usually positioned one-third of the way down from the beginning of the cut. If the bud shield is significantly narrower than the rootstock cut, line up one side exactly wrapping is extremely important in chip budding. If all exposed edges of the cut are not covered, the bud will dry out before it can take. Chip budding has become more popular over the past 5 years because of the availability of thin (2-mil) polyethylene tape as a wrapping material. This tape is wrapped to overlap all of the injury, including the bud, and forms a miniature lastic greenhouse over the healing graft.

c. Patch Budding

Patch Budding – probably the simplest to perform among the various methods of budding due to ease in removing or preparing rectangular patches of bark. It is widely used in plants with thick bark that can be easily separated from the wood. The method involves the complete removal of a rectangle-shaped patch of bark with the longer sides parallel to the axis of the stem of the rootstock. It is then replaced with a bud patch of the same size from a budstick. The patch of bud is cut from both the rootstock and the budstick by two parallel horizontal cuts either with one stroke of a double-bladed knife or two strokes when using a single-bladed knife. With vertical stroke of a knife, both horizontal cuts are connected at each side. The bud patch is carefully removed intact and inserted into the rootstock.

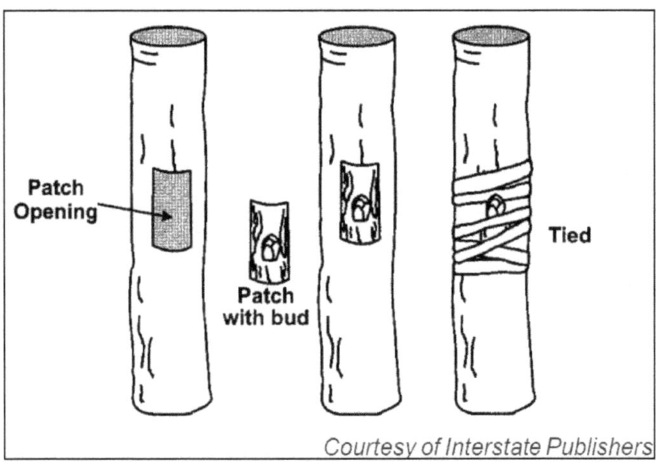

Courtesy of Interstate Publishers

d. Forkert Budding

A form of patch budding in which the patch of bark in the rootstock is retained. Incisions are made on the bark of the rootstock in the shape of "Ð" and pulled downward as a flap which is then used to cover the inserted bud patch. This flap is later removed to expose the bud. Both Patch and Forkert methods of budding follows the same procedure in the preparation of a bud patch.

e. Flute Budding

Similar to patch budding but the patch of bark that is removed from the stem of a rootstock almost completely encircles it except that there remains a narrow strip of bark (~1/8 the rootstock circumference) that connects the upper and the lower parts of the rootstock. The bud patch is prepared by two horizontal cuts about 2.5 cm apart (the same length as in the rootstock) in circular motion around the stem. The two cuts are then connected by a vertical cut and the patch of bark is separated intact from the wood. The circumference of the bud patch may be shortened by a vertical cut to fit into the rootstock. It is like that of a flute, this space is filled with scion.

f. Ring or Annular Budding

A method of budding the procedure of which is closely similar to the Flute budding technique. It involves the removal of a complete ring of bark from the rootstock without leaving a strip of bark that connects the upper and lower parts of the rootstock. As a result, a portion of the stem is girdled as if in preparation for marcotting. It is then replaced with a complete ring of bark with the same size from the budstick.

Advantages of Grafting and Budding

1. The plants which can not be propagated by other vegetative means *viz.* cutting, layers, or division can be multiplied, preserved and perpetuated by grafting and budding.
2. Grafting and budding can be very well adopted to convert inferior plant of established trees into superior one. Variety of the established plant can be changed by top working.
3. Root stocks influence size and vigour of tree and quality of fruits.
4. Root stocks impart disease resistance to the scion.
5. Some root stocks are tolerant to saline and alkaline conditions and high moisture contents of the soil.
6. Bridge grafting or buttress grafting helps to repair the damaged trunk or roots of the plant.
7. Special form of plant growth obtained by grafting and budding.
8. Novelty can be produced in nature by growing several types of flower or fruits on a single stock.
9. Early induction of flowers and fruits.
10. Budding and grafting are used for indexing the presence of virus disease. Virus susceptible stocks are used to detect the viruses. These are called indicator plants.

Disadvantages of Grafting and Budding

1. New varieties cannot be developed.
2. These are extensive methods of propagation. They require specialized skill.

3. The life span of grafted and budded plants is short as compared to seed propagated plants.
4. Spread of viral diseases may occur.

2.6 Garden Tools and Implements

A garden tool is any one or many tools made for gardens and gardening and overlaps with the range of tools made for agriculture and horticulture. Garden tools can also be hand tools and power tools.

Examples: Cultivators, String trimmer, Irrigation sprinklers, Hedge trimmers, Lawn aerators, Leaf sweepers, Trenchers, Leaf blowers, Chainsaws, Mini-tractors *etc.*

1. Garden Shovel

Digging is an essential part of gardening. Shovels are used to dig new beds, to carry soil and to fill holes. There are two common varieties of shovels used in gardening. The round-nosed shovel is used for digging holes for plants and for adding mulch, gravel or peat around plants. The flat blade of a square-nosed shovel allows gardeners to make clean, precise edges and lines; helps level soil; and carries gardening materials from one spot to another.

2. Garden Hoe

Gardeners use hoes to chop and remove weeds, for hilling soil around plants and to dig trenches for seed planting. The Dutch hoe is best used for breaking the soil surface and breaking up clumps of soil in large garden areas. Use a draw hoe or a short-handled onion hoe in smaller areas and in between plants. The Warren hoe is used to make furrows or holes for planting.

3. Spading Fork

The prongs of the spading fork allow it to easily open up rocky, clay-like soils in areas where a shovel may not do the job. Spading forks loosen solid soil and turn over soil in the garden. They are also helpful in breaking up large clumps of dirt and mud and in raking out stones and weeds from planting holes. Use a spading fork to spread organic matter, to move hay and manure, and to dig up bulbs and to lift ripe potatoes.

4. Wheel Barrow

Gardeners fill the wheelbarrow with gardening tools, plants, seed, soil, mulch and a multitude of items that they need to have near them when working in their garden. Older wheelbarrows were made of metal and were very sturdy and heavy, which made them somewhat difficult to maneuver up and down hills and over rocky terrain. Today, lighter plastic wheelbarrows are available. Plastic wheelbarrows are long-lasting, rustproof and are slightly lighter, allowing for easier movement.

5. Rakes

Moving or gathering up broken branches, leaves, sticks and twigs is made easier by using a rake. There are two types of rakes that are helpful in the garden. The leaf rake has long, plastic or flexible metal tines and is ideal for using to gather debris, leaves, grass clippings, gravel and mulch. A garden or bow rake has shorter, non-flexible tines that are attached the frame. Use a garden rake to spread and smooth out soil, mulch or other organic materials in garden beds.

6. Cultivator

This is a long handled tool with 3 angled prongs used to break down large clods of earth when preparing garden beds If you have a large plot it may be worth investing in a mechanical cultivator.

7. Draw Hoe

This is a long handled tool used for weeding using a chopping action and for drawing up soil around plants "earthing up". You can get also double hoes which incorporate a draw hoe and cultivator in one.

8. Spade

A long handled tool traditionally used for digging, shovelling soil and compost. They are available in various size.

9. Sprayers (Pressure)

Used for spraying plants with water or chemicals.

10. Watering Can

Used for watering smaller areas and containers garden fork.

11. Loppers

Used for taking smaller branches off trees. Often telescopic.

12. Pruners and Secateurs

Used for making sharp clean cuts on plant stems.

2.7 Manures and Fertilizers

Manure

Manure is organic matter, mostly derived from animal feces except in the case of green manure, which can be used as organic fertilizer in agriculture. Manures contribute to the fertility of the soil by adding organic matter and nutrients, such as nitrogen, that are utilised by bacteria, fungi and other organisms in the soil. Higher organisms then feed on the fungi and bacteria in a chain of life that comprises the soil food web.

Plant requires food/nutrients/elements for its growth and development which are absorbed through soil. The nutrient supplying sources are manures and fertilizers. Application of manures and fertilizers to the soil is one of the important factors which help in increasing the crop yield and to maintain the soil fertility. N, P and K are the 3 major elements required for the crop growth.

It is a well decomposed refuse from the stable and barn yards including both animal excreta and straw or other litter or the term manure implies to the any material with the exception of water which when added to the soil makes it productive and promotes plant growth.

Fertilizers

These are industrially manufactured chemicals containing plant nutrients or it is an artificial product containing the plant nutrients which when added to soil makes it productive and promotes plant growth. Fertilizer is any substance used to add nutrients to the soil to promote soil fertility and increase plant growth. Notice how nothing in there mentions the actual soil health; that's because not all fertilizers are made the same and not all are healthy for the soil.

2.8 Natural Organic Manures: Farm Yard Manure, Compost, Vermicompost and Biofertilizers

Organic manures not only supply nutrients to plants but also help in aeration of soil. They help excess water to drain out fast, but at the same time help to retain the moisture in the soil. By using the organic manures soil remains friable and thus encourages growth of hair roots. Organic manures are not water soluble; thus, the plants can not use these by themselves. Certain bacteria work upon these organic manures and gradually turn them to water soluble minerals. As this conversion process is slow and prolonged, the organic manures once applied will continue to provide nutrients to plants for quite some time.

Major sources of manures are:

1. Cattle shed wastes-dung, urine and slurry from biogas plants.
2. Human habitation wastes-night soil, human urine, town refuse, sewage, sludge and sullage.
3. Poultry, droppings of sheep and goat.
4. Slaughterhouse wastes-bone meal, meat meal, blood meal, horn and hoof meal, Fish wastes.
5. Byproducts of agro industries-oil cakes, bagasse and press mud, fruit and vegetable processing wastes *etc.*
6. Crop wastes-sugarcane trash, stubbles and other related material.
7. Water hyacinth, weeds and tank silt.
8. Green manure crops and green leaf manuring material.

Manures can also be grouped, into bulky organic manures and concentrated organic manures based on concentration of the nutrients.

a. Bulky Organic Manures

Bulky organic manures contain small percentage of nutrients and they are applied in large quantities. Farmyard manure (FYM), compost and green-manure are the most important and widely used bulky organic manures. Use of bulky organic manures has several advantages:

- They supply plant nutrients including micronutrients.
- They improve soil physical properties like structure, water holding capacity *etc.*
- They increase the availability of nutrients.
- Carbon dioxide released during decomposition acts as a CO_2 fertilizer.
- Plant parasitic nematodes and fungi are controlled to some extent by altering the balance of microorganisms in the soil.

b. Concentrated Organic Manures

Concentrated organic manures have higher nutrient content than bulky organic manure. The important concentrated organic manures are oilcakes, blood meal, fish manure *etc.* These are also known as organic nitrogen fertilizer. Before their organic nitrogen is used by the crops, it is converted through bacterial action into readily usable ammoniacal nitrogen and nitrate nitrogen. These organic fertilizers are, therefore, relatively slow acting, but they supply available nitrogen for a longer period.

Variety of manures and fertilizers available in the market are classified into two broad groups:

a) Natural or organic manures.

b) Artificial or chemical fertilizers.

Natural Organic Manures and Types

1. Farm Yard Manure

It is a mixture of cattle dung, urine, litter or bedding material, portion of fodder not consumed by cattle and domestic wastes like ashes *etc.* collected and dumped into a pit or a heap in the corner of the backyard. It is allowed to remain there and rot till it is taken out and applied to fields. Well rotten Farm Yard Manure contains 0.5. per cent N, 0.2 per cent P_2O_5 and 0.5 per cent K_2O.

Trenches of size 6 m to 7.5 m length, 1.5 m to 2.0 m width and 1.0 m deep are dug. All available litter and refuse is mixed with soil and spread in the shed so as to absorb urine. The next morning, urine soaked refuse along with dung is collected and placed in the trench. A section of the trench from one end should be taken up for filling with daily collection. When the section is filled up to a height of 45 cm to 60 cm above the ground level, the top of the heap is made into a dome and plastered with cow dung earth slurry. The process is continued and when the first trench is completely filled, second trench is prepared. The manure becomes ready for use in about four to five months after plastering. If urine is not collected in the bedding, it can be collected along with washings of the cattle shed in a cemented pit from which it is later added to the farmyard manure pit.

2. Compost

Well rotted plant and animal residue is called compost. Composting means rotting of plant and animal remains before applying in fields. The essential requirements of composting are air, moisture, optimum temperature and a small quantity of nitrogen. It is an activity of micro-organisms and same people recommend addition of suitably prepared inoculums to introduce micro-organisms for decomposing the material. Compost is organic matter that has been decomposed and recycled as a fertilizer and soil amendment. Compost is a key ingredient in organic farming. At the simplest level, the process of composting requires making a heap of wet organic matter known as green waste (leaves, food waste) and waiting for the materials to break down into humus after a period of weeks or months. Modern, methodical composting is a multi-step, closely monitored process with measured inputs of water, air, and carbon- and nitrogen-rich materials. The decomposition process is aided by shredding the plant matter, adding water and ensuring proper aeration by regularly turning the mixture. Worms and fungi further break up the material. Bacteria requiring oxygen to function (aerobic bacteria) and fungi manage the chemical process by converting the inputs into heat, carbon dioxide and ammonium.

a. Backyard Composting

- ☆ Select a dry, shady spot near a water source for your compost pile or bin.
- ☆ Add brown and green materials as they are collected, making sure larger pieces are chopped or shredded.
- ☆ Moisten dry materials as they are added.
- ☆ Once your compost pile is established, mix grass clippings and green waste into the pile and bury fruit and vegetable waste under 10 inches of compost material.
- ☆ Optional: Cover top of compost with a tarp to keep it moist. When the material at the bottom is dark and rich in color, your compost is ready to use. This usually takes anywhere between two months to two years.

b. Indoor Composting

If you do not have space for an outdoor compost pile, you can compost materials indoors using a special type of bin, which you can buy at a local hardware store, gardening supplies store, or make yourself. Remember to tend your pile and keep track of what you throw in. A properly managed compost bin will not attract pests or rodents and will not smell bad. Your compost should be ready in two to five weeks.

3. Vermicompost

It is the method of making compost with the use of earthworms, which generally live in soil, eat bio-mass and excrete it in digested form. This compost is generally called vermi-compost or wormi-compost. It is estimated that 1800 worms which is an ideal population for one sq. meter can feed up to 80 tones of humus per day.

Phases of Vermicomposting

Phase 1: Processing involving collection of wastes, shredding, mechanical separation of the metal, glass and ceramics and storage of organic wastes.

Phase 2: Pre digestion of organic waste for twenty days by heaping the material along with cattle dung slurry. This process partially digests the material and fit for earthworm consumption. Cattle dung and biogas slurry may be used after drying. Wet dung should not be used for vermicompost production.

Phase 3: Preparation of earthworm bed. A concrete base is required to put the waste for vermicompost preparation. Loose soil will allow the worms to go into soil and also while watering, all the dissolvable nutrients go into the soil along with water.

Phase 4: Collection of earthworm after vermicompost collection. Sieving the composted material to separate fully composted material. The partially composted material will be again put into vermicompost bed.

Phase 5: Storing the vermicompost in proper place to maintain moisture and allow the beneficial microorganisms to grow.

4. Green Manuring

Green manure crops are grown in the field itself either as a pure crop, or as an intercrop with the main crop, and buried in the same field. The most common green manure crops are sannhemp and guar. Tender green-twigs and leaves are collected from wastelands which are spread in the field and incorporated into the soil. Shrubs and trees are also cut and turned into the soil *e.g.* Shrubs like glyricidia, sesbania, karanj. Green manures are crops grown for the express purpose of plowing them in, thus increasing fertility through the incorporation of nutrients and organic matter into the soil. Leguminous plants such as clover are often used for this, as they fix nitrogen using *Rhizobium* bacteria in specialized nodes in the root structure.

Other types of plant matter used as manure include the contents of the rumens of slaughtered ruminants, spent grain (left over from brewingbeer) and seaweed

5. Biofertilizers

The biofertilizers containing biological nitrogen fixing organisms are of most important in agriculture.

Advantages of Biofertilizer

i) They help in the establishment and growth of crop plants and trees.

ii) They enhance biomass production and grain yields by 10-20 per cent.

iii) They are useful in sustainable agriculture.

iv) They are suitable in organic farming.

v) They play an important role in Agrotorestry/Silvi- pastaural system.

Types of Biofertilizers

a. Rhizobium

Most widely used biofertilizer is Rhizobium which colonizes the roots of specific legumes to form tumor like growths called root nodules. These nodules act as factories of ammonia production. The Rhizobium – legume association can fix up to 100-300 KG/N. in one crop season.

b. Azotobacter

Application of azotobacter has been found to increase yield of wheat, rice, maize, pearl-millet and sorghum by 0-30 per cent. Apart from nitrogen this organism is also capable of producing antifungal and antibacterial compounds, hormones.

c. Azospirillum

Certain micro-organisms like bacteria and blue green algae have the ability to use atmospheric nitrogen and transport this nutrient to the crop plants. Azospirillum is inoculated to maize, barley, oats, sorghum, pearlmillet and forage crops. It increases grain productivity of cereals by 5-20 per cent, of millets by 30 per cent and fodder by over 50 per cent.

d. Blue-green Algae

The utilization of blue green algae as a biofertilizer for rice is very promising. A judicious use of these algae could provide to the country's entire rice acreage as much nitrogen as obtained from 15-17 lakh tones of urea. Algae also helps to reduce soil alkalinity.

e. Azolla

A small floating water form Azolla is commonly seen in low land fields and in shallow fresh water bodies. These fern harbors a blue-green algae. Anabaena azollae. The Azolla – Anabaena association is a live floating nitrogen factory using energy from photosynthesis to fix atmospheric nitrogen accounting to 100-150 kg N/ha/year from about 40 – 60 tones of biomass.

f. Mycorrhizae

It is the symbiotic association of fungi with roots of vascular plants. It is useful in increasing phosphorus uptake *e.g.* in fruit crops like citrus and papaya.

Other Manures

1. Sheep and Goat Manure

The droppings of sheep and goats contain higher nutrients than farmyard manure and compost. On an average, the manure contains 3 per cent N, 1 per cent P_2O_5 and 2 per cent K_2O. It is applied to the field in two ways. The sweeping of sheep or goat sheds are placed in pits for decomposition and it is applied later to the field. The nutrients present in the urine are *wasted* in this method.

2. Poultry Manure

The excreta of birds ferment very quickly. If left exposed, 50 per cent of its nitrogen is lost within 30 days. Poultry manure contains higher nitrogen and phosphorus compared to other bulky organic manures. The average nutrient content is 3.03 per cent N; 2.63 per cent P_2O_5 and 1.4 per cent K_2O.

3. Oil Cakes

After oil is extracted from oilseeds, the remaining solid portion is dried as cake which can, be used as manure. The oil cakes are of two types:

- Edible oil cakes which can be safely fed to livestock; *e.g.*: Groundnut cake, Coconut cake *etc.*
- Non edible oil cakes which are not fit for feeding livestock; *e.g.*: Castor cake, Neem cake, Mahua cake *etc.*

2.9 Chemical Fertilizers–NPK

A fertilizer (American English) or fertiliser (British English) is any material of natural or synthetic origin (other than liming materials) that is applied to soils or to plant tissues (usually leaves) to supply one or more plant nutrients essential to the growth of plants.

Fertilizers enhance the growth of plants. This goal is met in two ways, the traditional one being additives that provide nutrients. The second mode by which some fertilizers act is to enhance the effectiveness of the soil by modifying its water retention and aeration.

A chemical fertilizer is a substance applied to soils or directly onto plants to provide nutrients optimal for their growth and development. The essential nutrients contained in these fertilizers are nitrogen, phosphorous, and potassium (NPK), as well as other nutritional substances in smaller amounts–all presented in a form that can easily be absorbed and metabolized by plants. Chemical fertilizers have become a staple in many yards and gardens, and can be a key component of a healthy lawn care routine. Read on for some basic information on chemical fertilizer and how you can use it most effectively in your yard.

Classification

Fertilizers are classified in several ways. They are classified according to whether they provide a single nutrient (*e.g.*, K, P, or N), in which case they are classified as "straight fertilizers." "Multinutrient fertilizers" (or "complex fertilizers") provide two or more nutrients, for example N and P. Fertilizers are also sometimes classified as inorganic (the topic of most of this article) versus organic. Inorganic fertilizers exclude carbon-containing materials except ureas. Organic fertilizers are usually (recycled) plant- or animal-derived matter. Inorganic are sometimes called synthetic fertilizers since various chemical treatments are required for their manufacture.

a. Single Nutrient ("Straight") Fertilizers

Single fertilizer is a fertilizer which contain single nutrient only. The main nitrogen-based straight fertilizer is ammonia or its solutions. Ammonium nitrate (NH_4NO_3) is also widely used. Urea is another popular source of nitrogen, having the advantage that it is solid and non-explosive, unlike ammonia and ammonium nitrate, respectively. The main straight phosphate fertilizers are the superphosphates. "Single superphosphate" (SSP) consists of 14–18 per cent P_2O_5, again in the form of $Ca(H_2PO_4)_2$, but also phosphogypsum ($CaSO_4 \cdot 2\ H_2O$). Triple superphosphate (TSP) typically consists of 44-48 per cent of P_2O_5 and no gypsum. A mixture of single superphosphate and triple superphosphate is called double superphosphate. More than 90 per cent of a typical superphosphate fertilizer is water-soluble.

b. Binary (NP, NK, PK) Fertilizers

This fertilizer provides two-component to the plants. Examples are NP,NK and PK. The main NP fertilizers are monoammonium phosphate (MAP) and diammonium phosphate (DAP). The active ingredient in MAP is $NH_4H_2PO_4$. The active ingredient in DAP is $(NH_4)_2HPO_4$. About 85 per cent of MAP and DAP fertilizers are soluble in water.

c. Multinutrient Fertilizers

These fertilizers are the most common. They consist of two or more nutrient components.

d. NPK Fertilizers

NPK fertilizers are three-component fertilizers providing Nitrogen, Phosphorus, and Potassium. NPK rating is a rating system describing the amount of nitrogen, phosphorus, and potassium in a fertilizer.

2.10 Time and Application of Manures and Fertilizers

The timing of application of manures and fertilizers must be adjusted in a way so as to make the nutrient available to the tree, when it needs them most. This period obviously, when the tree producing most of the growth flushes and, also during, flowering fruit set and fruit development. It appears desirable that manuring, should, be done before growth flush and as many times as there is a growth flush especially in light soils subject to leaching of plant food.

2.11 Different Methods of Fertilizer Application

It includes the methods like:

I) Broadcasting

Even and uniform spreading of manure or fertilizers by hand over the entire surface of field while cultivation or after the seed is sown in standing crop, termed as broad casting. Depending upon the time of fertilizer application, there are two types of broadcasting,

a) Broadcasting at planting

b) Top dressing

a) Broadcasting at Planting

Broadcasting of manure and fertilizers is done at planting or sowing of the crops with the following objectives:

1. To distribute the fertilizer evenly and to incorporate it with part of, or throughout the plough layer.
2. To apply larger quantities that can be safely applied at the time of planting and sowing.

b) Top Dressing

Spreading of fertilizers in the standing crop (after emergence of crop) is known as top-dressing. Generally, NO_3-N fertilizers are top dressed to the closely spaced crops like wheat and paddy.

II) Placement

In this, the fertilizers are placed in the soil irrespective of the position of seed, seedling or growing plant before or after sowing of the crops. It includes:

1. Plough Sole Placement

The fertilizer is placed in a continuous band on the bottom of the furrow during the process of ploughing. Each band is covered as the next furrow is turned.

2. Deep Placement or Sub-Surface Placement

In this method, fertilizers like Ammonium Sulphate and Urea, is placed in the reduction zone as in paddy fields, where it remains in ammonia form and is available to the crop during the active vegetative period. It ensures better distribution in the root zone, and prevents any loss by surface runoff. It is followed in different ways, depending upon local cultivation practices such as:

i) Irrigated Tracts

The fertilizer is applied under the plough furrow in the dry soil before flooding the land and making it ready for transplanting.

ii) Less Water Condition

Fertilizer is broadcasted before puddling which places it deep into the reduction zone.

3. Sub-soil Placement

This refers to the placement of fertilizers in the sub-soil with the help of heavy power machinery.

4. Localized Placement

It refers to the application of fertilizers into the soil close to the seed or plant. It is usually employed when relatively small quantities of fertilizers are to be applied.

Advantages

i) The roots of the young plant are assured of an adequate supply of nutrients.
ii) Promotes a rapid early growth.
iii) Make early Inter cultivation possible for better weed control.
iv) Reduces fixation of P and K.

It includes methods like:

a. Contact Placement or Combined Drilling or Drill Placement

It refers to the drilling of seed and fertilizer together while sowing. It places the seed and small quantities of fertilizers in the same row.

b. Band Placement

In this, fertilizer is placed in bands which may be continuous or discontinuous to the side of seedling, some distances away from it and either at level with the seed, above the seed level or below the seed level. There are two types of band placement

i. Hill Placement

When the plants are spaced 3 ft. or more on both sides, fertilizers are placed close to the plant in bands on one or both sides of the plants.

ii. Row Placement

When the seeds or plants are sown close together in a row, the fertilizer is put in continuous band on one or both sides of the one or both sides of the row by hand or a seed drill. For applying small amount of fertilizers, hill placement is usually most effective.

c. Pellet Application

In this method, fertilizer (nitrogenous fertilizers) is applied in the form of pellets

III. Side Dressing

Fertilizers are spread in between the rows or around the plants. It includes the application of Nitrogen containing fertilizers in between the rows by hand to broad row crops

IV. Injection

Injection methods are beneficial as they place liquid manure below the soil surface, eliminating both surface runoff on sloping soils, and volatilization of ammonia from the manure on any soil. Injection equipment can be added to liquid and slurry spreaders. Incorporation relies on nozzles close to the ground, either ahead or directly behind a tillage tool. Crop striping also appeared to be less of a problem when compared to machines with wider row spaces.

V. Foliar Sprays

Water-soluble fertilizer can be applied with spray cans or applied with sprinklers or furrow irrigation. Small amounts of liquid fertilizer applied to young vegetable plants at the time of transplanting is called a starter solution. Foliar spray fertilizer is a good way to supplement the nutritional needs of your plants. There are various types of foliar spraying options available to the home gardener, so finding a recipe or suitable solution to accommodate your needs should be easy. Foliar plant spray involves applying fertilizer directly to a plant's leaves as opposed to putting it in the soil. Foliar spray fertilizer is an excellent short-term solution for plants experiencing stress. However, it is always best to build up your soil with plenty of organic matter.

Chapter 3

GARDEN

3.1 Garden and its Components

A garden is a planned space, usually outdoors, set aside for the display, cultivation, and enjoyment of plants and other forms of nature. The garden can incorporate both natural and man-made materials. Gardening is the activity of growing and maintaining the garden. This work is done by an amateur or professional is called as gardener. A gardener might also work in a non-garden setting, such as a park, a roadside embankment, or other public space. Landscape architecture is a related professional activity with landscape architects tending to specialize in design for public and corporate clients.

The major components:

- Softscape - trees, plants, grass, mulches
- Hardscape - sidewalks, walls and fences, pavers, rocks, decks and patios
- Décor - water features, statuary, tree hangers, pottery, lighting
- Supporting Features - birdhouses, beehives, feeders, tree houses, trellises

a. Softscape

Softscape is any component of a garden that is or used to be alive. Ground covers, bushes and trees are major softscaping components.

b. Hardscape

Hardscape refers to items that are permanent and do not break down. It has many uses and comes in forms adapted to those uses. It contains:

- Edgings between plants and grass

- ☆ Walls and fences
- ☆ Walkways and driveways
- ☆ Patios and deckings
- ☆ Structures like gazebos
- ☆ Garden benches or other seating
- ☆ Ponds and pools

c. Landscape Garden Decor

Decor is what is used to enhance a garden, once the main features have been installed. Generally a garden will not need much in the way of décor.

d. Supporting Landscape Features

There is something endearing and calming about gardens. Aside from contributing to a greener earth, a small patch of garden brings you in tune with nature.

Principles of Gardening

1. Unity

Landscape elements such as plants and flowers, cast or carved stone fountains, garden set and other garden accessories must complement each other. Even an immaculately kept garden will not be as restful to look at if one or two components are out of sync.

2. Balance

A garden is either symmetrical or asymmetrical. Symmetrical balance is easier to achieve for it is merely dividing an area into two equal parts and designing them exactly the same way. Asymmetrical is more interesting but harder to achieve.

3. Color

A good mix of colors is good for the garden.

4. Proportion

Each and every component in the garden must be in direct proportion with the rest of the landscape elements. A rather large water feature has no place in a tight garden area.

5. Line

A line pertains to the walkways and pathways in a garden. These paths make the flow from one area of the garden to the other flawless, seamless and effortless. This means that the path should be clear from any obstruction or structure unless it's the end of the path.

3.2 Components of Garden - Living and Nonliving Components

1. Lawn

Lawn is the green carpet for a landscape. It provides a perfect setting for a flowerbeds or shrubbery or specimen tree.

2. Shrub and Shrubbery

Shrubs are small plants having a varying length of 1 to 5 mts. A shrubbery is a border planted with a different kinds of shrubs.

3. Climbers and Creepers

Creepers are those plants which are unable to climb vertically on their own because of their week. Climbers and Creepers are used to grow against or over walls, trellises, arches, pergolas, pillars or any large trees.

4. Flower Beds and Borders

Flower borders are continuous beds of more length than width containing plants of heterogeneous character as flower beds which are composed of plants of one kind only. These borders can be on the sides of path,walks or in front of shrubberies and trellises with climbers.

5. Ornamental Hedges

Ornamental hedges planted in garden with attractive foliage or flowering shrubs. These are pruned to maintain the height of 50-65cm.This will help to divide the garden into number of parts each will have its own distinct features.

6. Rockery

This is elevated structure resembling a miniature mountain range or slope of a hill with few dominant peaks or valleys. It contains some of the Cacti and Succulents used in rockery are *Agave, Aloe vera, Sedum, Yucca.* Ferns-*Nephrodium, polypodium.* Flowering-*Vinca rosea, Verbena.*

7. Carpet Beds

These are designs or letters formed by plants. In a large public garden close growing plants like *Verbena or Alternanthera* are used to form certain designs or letters this form carpet beds. Carpet beds require constant attention and need not allowing them to over grow.

8. Topiary

Topiary is process of making top works in plants. Certain plants are often trimmed to shapes of animals or birds *etc.* Shrubs are well suited for bending and withstand frequent trimming for developing topiary. Shrubs like *Casuariana, Bougainvillea, Cupressus* are suitable for Topiary work.

9. Trophy

Arrangement of potted colorful foliage or flowering shrubs and flowering annuals or herbaceous perennials around a tree or any central objects such as statue.

10. Edges

These are materials of any description which is used in gardens for dividing beds, borders from the roads, walks or paths demarcating spaces allotted for particular purposes as flower beds.

11. Trees

Trees are form main frame work of the garden. Generally planted along the boundaries.

12. Garden Adornments

There are several adornments and accessories such as fountains, garden seats, statues, ornamental posts and pillars, terllises, hanging baskets, tubs, vases and urns with plants make the garden more enjoyable.

13. Sunken Garden

These are the natural depression of deepness of a garden. It breaks the monotony of a flat ground of the garden. These garden goes down through a series of terraces to a small pool or a fountain at a bottom.

14. Arch and Pergola

Arches are generally erected over walks, usually at the entrance and are two meters in height. Pergolas are series of arches connected over a walks

15. Walks, Paths, Drive, Road

These are way for travelling in and around the garden. Paths may be made of earth,brick,concrete or paved. Paved paths are effective in formal garden. Paving can be done flat stones or concrete slabs or bricks

16. Garden Ponds

Self-enclosed systems can include waterfalls, bridges, fountains, and even fish.

17. Lighting

These are different sources for light, which provide beauty to garden during day and night

18. Fences

These are protective covering to garden. Not only do fences provide security and privacy, they can substantially improve the attractiveness of a landscape.

3.3 Lawns and Landscaping: Trees, Shrubs and Shrubberies, Climbers and Creepers

A lawn is an area of soil-covered land planted with grasses or other durable plants such as clover which are maintained at a short height with a lawnmower and used for aesthetic and recreational purposes. Common characteristics of a lawn are that it is composed only of grass species, it is subject to weed and pest control, it is subject to practices aimed at maintaining its green color (*e.g.*, watering), and it is regularly mowed to ensure an acceptable length, although these characteristics are not binding as a definition. Lawns are used around houses, apartments, commercial buildings and offices.

Landscaping refers to any activity that modifies the visible features of an area of land, which includes the proper planning of the following components

1. Living elements, such as flora or fauna; or what is commonly called gardening, the art and craft of growing plants with a goal of creating a beautiful environment within the landscape.
2. Natural elements such as landforms, terrain shape and elevation, or bodies of water; and
3. Abstract elements such as the weather and lighting condition.

Plants and its Classification

Based on growth habit plants are broadly categorized into three. They are as:

a. Trees

A tree is a perennial plant with an elongated stem or trunk, supporting branches and leaves in most species. In some usages, the definition of a tree may be narrower, including only woody plants with secondary growth, plants that are usable as lumber or plants above a specified height. A tree typically has many secondary branches supported clear of the ground by the trunk.

e.g. Pine, Polyalthia, Cherry, Oak, Cypress, Fig *etc.*

b. Shrubs

A shrub or bush is a small to medium-sized woody plant. Unlike herbs, shrubs have persistent woody stems above the ground. They are distinguished from trees by their multiple stems and shorter height, and are usually under 6 m (20 ft) tall. Plants of many species may grow either into shrubs or trees, depending on their growing conditions. Small, low shrubs, generally less than 2 m (6.6 ft) tall, such as lavender, periwinkle and most small garden varieties of roses, are often termed "subshrubs". Shrubs are a key foundation planting for many gardens. They offer structure and organizing points; many also supply year-round color, as well as food and shelter for wildlife.

Importance of Shrubs in Garden

1. Being permanent, they form part of the frame work of the garden.
2. They form the chief features of landscape, gardenings placed in front of tall trees and joining the spacious lawn *etc.*
3. Shrubs which are amenable for frequent training are chosen for topiary work
4. Tall growing shrubs often serve as screen
5. They are useful as a single specimen in the lawn
6. They can be trained to form standards *i.e.*, trained to single stem and allowed to branch out and form a handsome head only above a particular height

   e.g. Gardenia, Oleander, Spruce, Balsam, Ixora, Murrya *etc.*

c. Shrubberies

A shrubbery is a wide border to a garden where shrubs are thickly planted, or a similar larger area with a path winding through it. A singular shrub is also known as a bush. A shrubbery was a collection of hardy shrubs, quite distinct from a flower garden, which was a cutting garden to supply flowers in the house

e.g. Gardenia, Oleander, Spruce and Balsam *etc.*

d. Creepers

Creeping plants or "creepers" are generally considered to be small, viny plants that grow close to the ground. They are sometimes called "procumbent," as well. In cases where their vines are long enough and you wish to have them climb a structure, you need to guide them (train them) and secure them to a support if they are to achieve much height at all. In this sense, they differ from "climbers," which are another class of vine.

e.g. Berry plants (strawberry, cranberry), Forage plants (white clover), and food plants (sweet potato) *etc.*

e. Climbers

Climbers are a versatile group of plants - use climbing plants to cover fences, walls, trellis, arches or obelisks! You can also train certain shrubs as 'wall shrubs' for the same effect. Climbers are great for screening unsightly areas of the garden, brightening up bare walls and adding height to your borders.

Uses of Climbers in Garden

1. Certain climbers are grown in gardens for their attractive foliage. *e.g. Asparagus spregeri, Ficus repens, Hedera helix, Scindapsus aureus.*
2. Some light climbers can be trained as 'screens' in gardens. *e.g. Bignonia venusta, Jacquemontia violaceae, Passiflora edulis.*

3. Climbers like *Allamanda, Antigonon, Aristolochia elegans, Solanum seaforthianum* can be used on arches, bowers and pergolas.
4. Heavy climbers like Bougainvillea, Scindapsus, *Petrea volubilis* can be trained over strong pergolas or on trees which look very attractive

   e.g. Morning glory, Ipomoea species, Passiflora, Clematis species, Climbing rose *etc.*

3.4 Flower Beds and Borders

Annuals and herbaceous perennials are grown in flower beds to provide massing effect of different colours. Borders are continuous beds of more length than width containing plants of one kind only.

a. Hedges

A "hedge" is a wall composed of plants. Some are purely decorative, while others serve primarily a practical function. Hedge plants used decoratively are often trimmed to precise sizes and shapes and include evergreen and deciduous shrubs. Hedges used to separate a road from adjoining fields or one field from another, and of sufficient age to incorporate larger trees, are known as hedgerows.

e.g. Acalypha sp., *Barleria* sp., *Bougainvillea* sp., *Cupressus macrocarpa, Duranta plumieri, Hamelia patens, Hibiscus rosa-sinensis, Lantana camera.*

b. Edges

The very name, "edging plants" indicates how these plants are used or are grown along the edges (borders) of various features in the landscape, for example, the edges of driveways. Edges are plants which are employed in gardens for dividing beds, borders, roads, walks or path of demarcating spaces allotted for particular purpose, as flower beds. The edging plants should be perennial, hardy, easily propagated and should have lasting foliage or flower or both. The height of an edge should be about 15 to 20 cm. Constant trimming is necessary to maintain proper shape and height.

e.g. Alternanthera, Coleus sp., *Eupatorium cannabinus, Iresine* sp., *Santolina chamaecyparissus.*

c. Drives, Roads, Walks and Paths

Drives, Roads Walks and Paths are prime features of any garden. Providing access and parking, drives can be constructed from a wide variety of materials including permeable paving blocks, gravel, ground reinforcement mesh or units to comply with current planning drainage requirements. Paths give a defined progressive route around, through and linking the garden spaces. They can have both an aesthetic and a practical use.

d. Carpet Beds

The art of growing ground cover plants closely and trimming them to a design or alphabetical letters is called a carpet bed. Colourful foliage as edge plants is found to be more suitable for this purpose'.

e.g. Alternenthera.

e. Topiary

Topiary is the horticultural practice of training perennial plants by clipping the foliage and twigs of trees, shrubs and subshrubs to develop and maintain clearly defined shapes, whether geometric or fanciful. The term also refers to plants which have been shaped in this way. As an art form it is a type of living sculpture. The word derives from the Latin word for an ornamental landscape gardener, topiarius, a creator of topia or "places", a Greek word that Romans also applied to fictive indoor landscapes executed in fresco.

e.g. Cupressus macrocarpa, Pinus patula Casuarina sp., *Caesalphinia coriari, Bougainvillea* sp., *Jaminum* sp., *Thuja species, Laurus Buxus sempervirens etc.*

f. Trophy

It is the arrangement of colourful potted plants in differ-ent tiers around a central object which may be a tree trunk, lamp post or a pillar.

g. Rockery or Rock Garden

A rockery or rock garden is the arrangement of rocks with plants grown in the crevices. A rock garden, also known as a rockery or an alpine garden, is a small field or plot of ground designed to feature and emphasize a variety of rocks, stones, and boulders. In plains, on the sunny side some of the cacti and succulents; *Lantana, Setcreasea, Verbena, etc.* can be grown successfully. Ferns and some indoor plants also look natural on the rockery slopes in shade.

3.5 Conservatory or Greenhouse, Indoor Garden, Roof Garden

Conservatory or Greenhouse

A conservatory is a building or room having glass or tarpaulin roofing and walls used as a greenhouse or a sunroom. Conservatories originated in the 16th century when wealthy landowners sought to cultivate citrus fruits such as lemons and oranges that began to appear on their dinner tables brought by traders from warmer regions. Shade loving plants are mainly grown in pots which are highly priced for the ornamental foliage. It is also called as greenhouse and glass house. It helps to create a special environment with cool, airy, humid and marshy soil for shade loving, moisture loving and water loving plants. The root of such structure is made of wooden reapers, angle irons and concrete poles and covered with mesh or natural creepers. A tank or water trough is constructed at the center to increase humidity and few benches are provided for the display of plants. Such structures can be erected under a partial shade preferably.

A greenhouse (also called a glasshouse, or, if with sufficient heating, a hothouse) is a structure with walls and roof made chiefly of transparent material, such as glass, in which plants requiring regulated climatic conditions are grown. These structures range in size from small sheds to industrial-sized buildings. A miniature greenhouse is known as a cold frame. The interior of a greenhouse exposed to sunlight becomes significantly warmer than the external ambient temperature, protecting its contents in cold weather.

Greenhouses allow for greater control over the growing environment of plants. Depending upon the technical specification of a greenhouse, key factors which may be controlled include temperature, levels of light and shade, irrigation, fertilizer application, and atmospheric humidity. Greenhouses may be used to overcome shortcomings in the growing qualities of a piece of land, such as a short growing season or poor light levels, and they can thereby improve food production in marginal environments. Greenhouses in hot, dry climates used specifically to provide shade are sometimes called "shadehouses".

Greenhouses are often used for growing flowers, vegetables, fruits, and transplants. Special greenhouse varieties of certain crops, such as tomatoes, are generally used for commercial production. Many vegetables and flowers can be grown in greenhouses in late winter and early spring, and then transplanted outside as the weather warms. Bumblebees are the pollinators of choice for most pollination although other types of bees have been used, as well as artificial pollination. Hydroponics can be used to make the most use of the interior space.

3.6 What is Indoor Gardening?

Indoor gardening is a process of creating an artificial garden environment inside a house. It is beneficial for those living in homes or apartments where there is no room to garden. It's also perfect for gardeners who would still like to have fresh produce during the cold winter months.

Material Required for Indoor Gardening

Lighting

For those who have to use artificial lighting, the good news is that there are endless options for grow lights, need about 100 watts of lighting for every 18"x18" patch of garden. If you looking for cheap grow lighting, the best way to go is to purchase spiral CFL bulbs from your local superstore.

General Supplies

For soil gardeners, you'll want to pick up a premium organic potting soil and either purchase or recycle containers for your plants to grow in.

3.7 Roof Garden

A roof garden is a garden on the roof of a building. Besides the decorative benefit, roof plantings may provide food, temperature control, hydrological benefits, architectural enhancement, habitats or corridors or wildlife, recreational opportunities, and in large scale it may even have ecological benefits. The practice of cultivating food on the rooftop of buildings is sometimes referred to as rooftop farming. Rooftop farming is usually done using green roof, hydroponics, aeroponics or air-dynaponics systems or container gardens.

Roof gardens are most often found in urban environments. Plants have the ability to reduce the overall heat absorption of the building which then reduces energy consumption. "The primary cause of heat build-up in cities is insolation, the absorption of solar radiation by roads and buildings in the city and the storage of this heat in the building material and its subsequent re-radiation.

Benefits of Roof Garden

- They make use of unused or underused space
- A garden beautifies an eyesore
- They can provide privacy
- They can be extremely environmentally friendly
- There is usually good sun exposure

3.8 Bonsai

The word "Bon-sai" (often misspelled as bonzai or banzai) is a Japanese term which, literally translated, means "planted in a container". This art form is derived from an ancient Chinese horticultural practice, part of which was then redeveloped under the influence of Japanese Zen Buddhism.

What is a Bonsai Tree?

Bonsai can be explained as, "Bon" [left character] is a dish or thin bowl ("a modified vessel which has been divided or cut down from a deeper form")."Sai" [right character] is a tree or other growing plant which is planted – "planted," as would be a halberd or spear or pike stuck into the ground."Bonsai" thus means or denotes "a tree which is planted in a shallow container".

Different Types and Styles

a. Trunk and Bark Surface

The type of trunk and bark on the bonsai play an important role in its style. There are many terms used to describe different bark and trunk forms – for example, a bonsai with a twisted trunk is known in the traditional Japanese as a *nebikan*, while those species with dead branches or trunks are known as deadwood bonsais.

b. Trunk Orientation

If the trunk of the tree has its apex directly above the base of the trunk (where it enters into the potting soil), it is known as an upright style. (*Chokkan* and *Moyogi* in Japanese). If the trunk is slightly slanted, it is known as an informal upright, and if the branches hang lower than the base of the trunk, the style is known as cascade. Branches at the same level as the base of the trunk are known as the semi-cascade style.

c. Root Status

Most bonsai trees are planted directly into soil; however there are certain bonsai species that are planted with their roots over rocks. This root-over-rock style is known as *deshojo*, and a similar style in which the entire tree is rooted within a rock is known as *ishizuki*.

d. Number of Trunks

Most bonsais consist of just a single tree with a single trunk. However, there are also specialised styles for those bonsais with a number of trunks, or multiple trees, also known as forest style bonsais. There are style categories for trees with multiple trunks coming off a single root, as well as for completely separate trees in one tray known as forest style. In addition, the configuration of multiple trunks and roots also plays a role in the style categorization – multiple trunks growing from a mass of interconnected roots are known as raft or sinuous style, and the Japanese terms *yose-ue* is used to describe any number of multiple, separate bonsais in one tray. These classification systems of bonsai styles can be quite confusing, particularly as the styles are not mutually exclusive – a single bonsai tree can fit into a number of different categories. In these situations, the bonsai is generally described by the style which is most prominent.

e. Indoor and Outdoor Bonsai

Within all of these classification systems, there are two large groups of Bonsais

- indoor and outdoor plants. Within these large groups, there are numerous different varieties that can suit the desires of the grower.

Bonsai Styles

1. Coniferous

Coniferous bonsai is one of the more challenging forms of this beautiful art of miniatures. Conifers are slow-growing and require a great deal of patience, study and care.

2. Deciduous

Unlike Evergreens, which keep their leaves all year round, deciduous trees lose their leaves during the winter months. The term "*deciduous*" actually means, "*falling off at maturity*".

3. Evergreen

Evergreens are often used to create a miniature tree that can be enjoyed indoors or out. Some of the more common species of evergreen bonsai include spruce, cedar, and redwood to name a few.

4. Flowering

Flowering bonsai plants require regular watering, feeding, training and pruning like other ornamental and flowering plants. They're grown in small containers and are trained with wire coils to induce their desired shape and structure.

5. Tropical

With tropical varieties, you'll find plants that enjoy temperatures that are higher than 60 degrees Fahrenheit, which are ideal for indoor maintenance.

6. Shohin

Of the many types of bonsai, *Shohin* is among the most challenging and most rewarding. The word Shohin literally means "*small goods*". Trees trained to the Shohin size are no more than 12 inches tall or wide, and can be much smaller. A good rule of thumb is that the potted tree should be small enough to be held on the palm of one's hand.

Chapter 4

FLOWER, FRUIT AND VEGETABLES

4.1 Flower Arrangement

Flower arrangement may be defined as the art of organizing and grouping together plant materials (flowers, foliage, twigs, *etc.*) to achieve harmony of form, colour, and texture, thereby adding cheer, life, and beauty to the surroundings. Flower arrangement is an organization of design and color towards creating an ambience using flowers, foliage and other floral accessories. Flower arrangement is not an art to be reserved only for special occasions.

Basic Ingredients of Flower Arrangement

1. Mechanics

These are items used to keep flowers, foliage, and stems in place within the container. Mechanics must be fixed securely and should be hidden from view.

Examples for mechanics are – florists' foam (oasis), pin holders (Japanese term – kenzan), chicken wire, prong, adhesive clay and tape, florist cone.

2. Equipments

This includes tools used to ensure that a satisfactory arrangement of plant material is created within the container.

Examples – bucket, scissors, knife, watering can, mister, wire cutter, cocktail sticks, turn, wire, floral tape, candle holder, cut flower preservatives, and secateurs.

3. Containers

These are receptacles that hold the flower arrangement. They may or may not be hidden by the plant material. The container must be waterproof and neutral colours such as soft grey, dull brown, off-white, or earth colours are most suitable because they are inconspicuous and do not detract attention from flowers displayed.

Example – vases and jugs, basket, bowls and trays, wreath frame *etc.*

4. Bases

An object that is placed underneath the container to protect the surface of the support and/or to add to the beauty of the display is called a base.

Example – table mat, tree section, wood base, stone base, and oriental base.

5. Support

This refers to the structure on which the container stands.

Example – tables, sideboards, alcoves, and shelves.

6. Plant Materials

These can be divided into 3 basic types:

a) **Flowers (Dominant/Focal/Point Material)**: This consists of bold flowers or clusters of small showy blooms. The dominant material provides a centre of interest.

b) **Fillers (Secondary Material)**: This consists of smaller flowers and all sorts of leaves and foliage that are used to cover the mechanics and edges of the container and also provide added interest and colour to the display.

c) **Foliages (Line material)**: This consists of tall stems, flowering spikes, or bold leaves that are used to create the basic framework or skeleton.

7. Accessories

These are non-plant materials included in or placed alongside the arrangement. Their purpose in generally decorative but could be functional at times.

Example – miniature dolls, hats, ribbons, beads, painted wire, wooden fruit shapes, silk flowers and foliage, candles, driftwood, shells, idols, interesting pebbles *etc.*

Principles or Design of Flower Arrangement

1. Scale

Scale is easy to understand as we can all recognize when small flowers look wrong in a large container, *etc.* However size is relative – an object seen by itself is not big or small unless it is seen next to another object for comparison. If the difference is great then objects do not go well together.

2. Proportion

Good proportion refers to pleasing amount of things and again it a matter of relationships. The same amounts of material that appears too much for one container may seem correct for another. A number of arrangements can be made for a room and all may be in scale with their setting but the number of arrangements may be too many, so the proportion of arrangements to the room is not pleasing.

3. Balance

Here, physical as well as visual balance needs to be considered. Physical balance is vital for any arrangement. The more one-sided the display, the heavier the container should be filled with sand and gravel can be added. Visual balance, calls for the arrangement to look stable even if it is one-sided. Visual balance depends on the eye being attracted to both sides of the design equally by the use of different colours, shapes and textures.

4. Rhythm or Movement

This involves using techniques and materials that guide the eye from one part of the display to another. Rhythm can be in colour and size. Rhythm in flower arrangements may be achieved by using curved stems, hiding all or part of any tall, straight stems, placing flowers 'in and out' through the arrangement, having flowers at various stages of development in the arrangement, using foliage of various sizes and contrasting shapes, having an irregular line of various- sized blooms.

5. Contrast

Contrast can be created in shape – by turning the flowers to different ways when all round flowers are used. Contrast can be achieved by introducing line plant material. Strong contrast in textures can be used for interest. Very strong contrasts should be avoided as too much contrast may upset the unity of the arrangement.

6. Emphasis/Dominance

This involves having one or more areas in the arrangement to which the eye is drawn and on which it rests for a short time. This point is known as a 'focal point' or 'centre of interest'. The usual methods to achieved by a small group of bold flowers (dominant material), use an unusual container. Use striking foliage, have sufficient plain background.

7. Harmony/Ynity

In a pleasing flower arrangement, the plant material, container, base, accessory and setting should all be in harmony. Similarity in appearances between materials help s give repetition and a feeling of harmony, *e.g.*, the curve of a piece of driftwood, *etc.*

Types of Flower Arrangements

Classification I

1. Floral Bouquets

Floral bouquets are simple and easier to assemble than a formal flower arrangement. Bouquets made of roses are the most popular.

2. Table Centerpieces

Centerpieces are nothing but the normal floral arrangements, designed to be placed on tables when dinners are hosted on such occasions like Thanksgiving day, Christmas, *etc.*

3. Floral Baskets

Floral baskets are flower arrangements done in baskets of varied depths. Flowers for the floral baskets should therefore be chosen on the basis of the depth of the basket.

4. Floral Wreaths

A floral arrangement in the form of a circular garland, usually woven of flowers and foliage, that traditionally indicates honor or celebration.

Classification II

1. Oriental Flower Arrangements

The emphasis is on the *Lines* in the floral arrangement.

2. Traditional/Western Flower Arrangements

The emphasis is on the *Use of many flowers as a mass* in the floral arrangement.

3. Modern Flower Arrangements

There are no rules at all. These floral arrangements are based on the taste of the floral designer. But, it is more close to Oriental than Traditional style of arranging flowers in its inclination towards importance to lines.

Classification III

1. All-round Arrangements

This arrangement is designed to be seen from all sides and is therefore chosen for a table or a room centerpiece.

2. Facing Arrangement/Flat-Back Arrangement

It is designed to be seen only from the front and perhaps from the sides. It is therefore chosen for placement on a shelf or sideboard.

3. Mass Style

Little or no space is enclosed within the boundary of the arrangement. This style is originated in Europe.

4. Line Style

In this style, open spaces within the boundary of the arrangement are the main feature. Most of the display is line material. The basic feature of a line design is limited use of plant material with support often provided by a pin holder.

5. Triangular Shape

It is a popular shape for symmetrical arrangements. The first step is to establish lines of height and width, usually with flowers or foliage of finer form or paler colour. The next step is to establish a focal point of interest with large or darker-coloured flowers. Fill in with flowers of varied stem lengths, grouping colours.

a. Left triangle-made in a shallow container with consecutive stem along the left side.

b. Right triangle - like the left triangle arrangement, but the tallest stem on the right side of the container with consecutive stem.

6. Vertical Line

A very tall arrangement placed in long and cylindrical flower vase using a very long stemmed flower like torch lilies.

7. Line Mass Style

In this style, some open space is present within the boundary of the arrangement.

a. Circular shape - or round shape, is an arrangement in which flowers are arranged in circular designs.

b. Crescent shape - it is asymmetrical and formal arrangement which requires more skill and experience.

c. Fan shape - the fan or horizontal shape is a good line to follow when designing flowers for the centre of the table. It is a low arrangement, symmetrical and thus attractive from every angle.

d. Hogarth or 'S' shape - this style was pioneered by an 18th century painter, William Hogarth. This is a very graceful and easier to make arrangement when curved branches and pliable stems are used. After establishing the S shape with these, flowers are filled in at the centre and just above and below the rim of the tall container.

8. Miscellaneous Style

a. Parallel style/European style

b. Landscape style

c. Foliage arrangement
d. Dried flower arrangement

Classification IV

1. Formal Arrangement

This is symmetrical and precise.

2. Semi-Formal Arrangement

This is more or less symmetrical in outline.

3. Informal Arrangement

This is asymmetrical and 'free'.

4. Modern or Abstract or Free-Style Arrangement

these have no fixed rules for correct proportions. These arrangements do not have a definite geometric outline; instead the emphasis is on line and space. The individual beauty of each piece of plant material is emphasized instead of the beauty of an outline shape or a mass

Classification V

Eight Basic Flower Arranging Designs

Though so similar in use, artificial flowers are an entirely different art form. The line, focal, filler technique is an excellent way to design artificial and dried flower arrangements.

a. Horizontal Arrangements

1. Using a relatively shallow container, anchor foam with a lot of glue or use anchor pins, and position sprays of line flowers to establish the shape of the design.

2. Insert focal flowers in the middle so they gently droop over the lip of the container on both sides, reach towards the line material and extend on either side of the middle. Leave room for filler flowers.

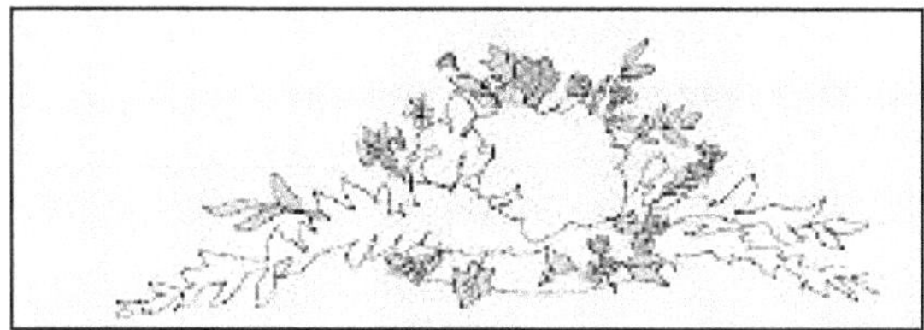

3. Fillin and around focal area with filler flowers and foliage.

b. Vertical Arrangements

1. Wedge or secure foam in a vase with hot glue. Cut the stems of the tallest flowers or leaves to reach three or four times the height of the vase.
2. Place the focal flowers vertically within the diameter of the vase.
3. Fill in the areas as needed with filler flowers.

c. Triangular Arrangements

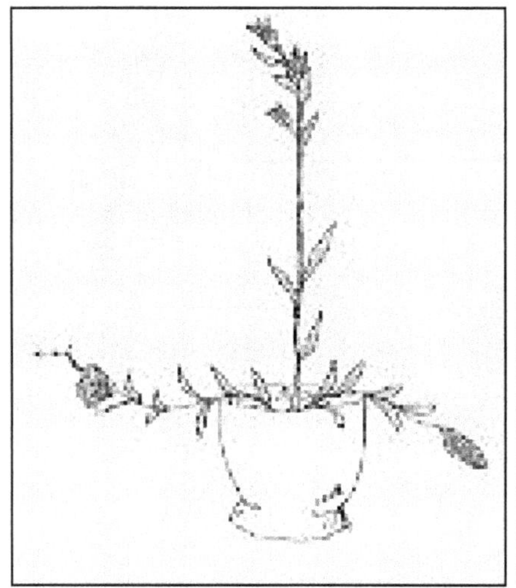

1. Secure floral foam. Determine the vertical height and horizontal width with the smallest line flowers and/or leaves. Make the height higher than the width.

2. Position the largest focal flowers in the heart of the arrangement and slightly lower to give weight and balance.

3. Fill in with the filler flowers and foliage keeping within the triangular shape.

d. Crescent Arrangements

1. Secure foam in container. Determine length of crescent and insert curved line flowers or leaves to follow the crescent form. Angle the shape to balance in the container.

2. Insert the focal flowers fairly low in the bowl to achieve balance, stability and depth.

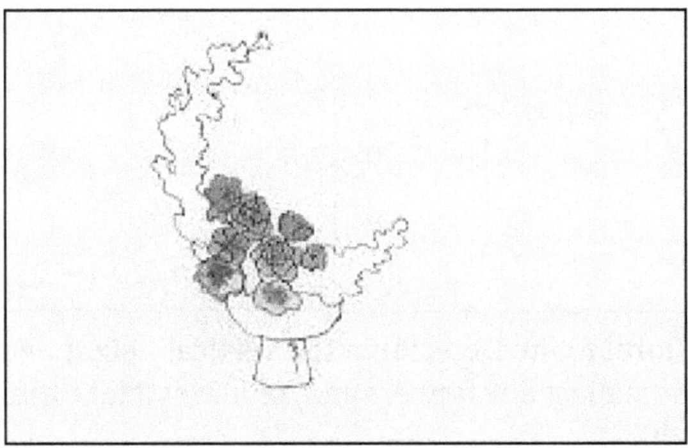

3. Fill in around the focal flowers with smaller flowers and foliage. Place wisps of filler flowers that gracefully taper off the ends.

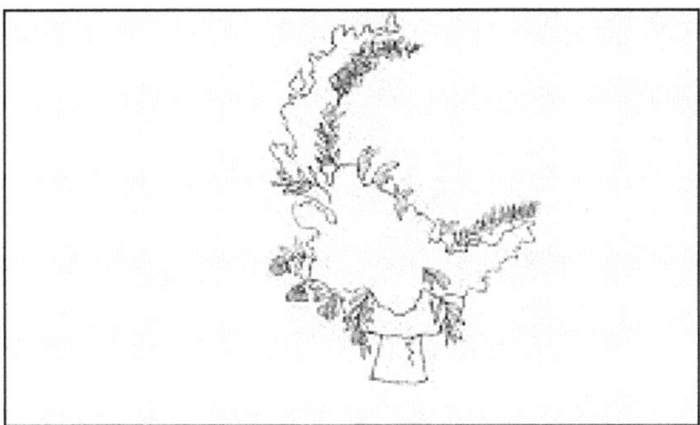

e. Oval Arrangements

1. Secure the floral foam. Determine the height with line flowers, then frame in the outer edges of the oval shape with light colored flowers and foliage.

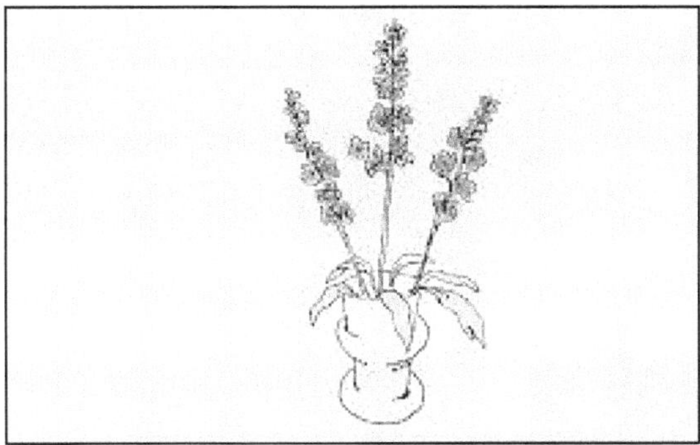

2. Place the largest, strongest or brightest flowers in the focal area.

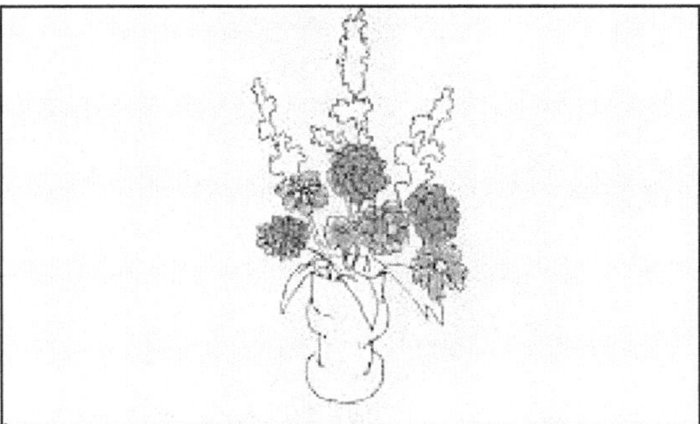

3. Fill in around the larger flowers and leaves with the filler flowers.

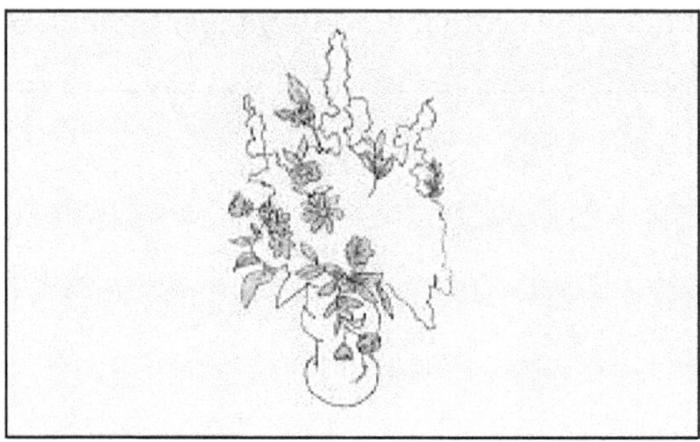

f. Minimal Arrangement

1. Adhere foam to container. Insert vertical line flowers to determine the height, and secure the horizontal line flowers to give the basic outline of the arrangement.

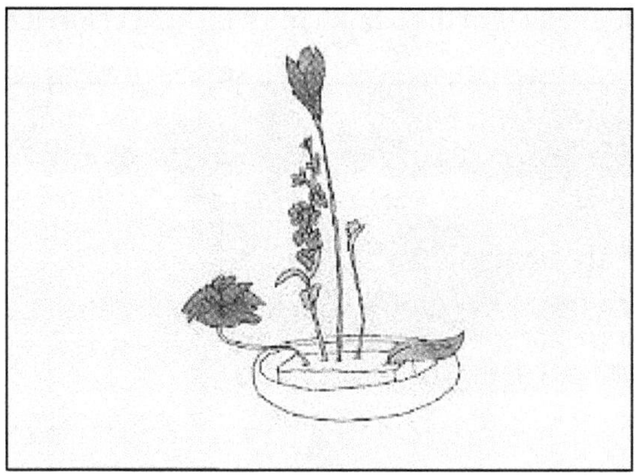

2. Place the focal flowers.

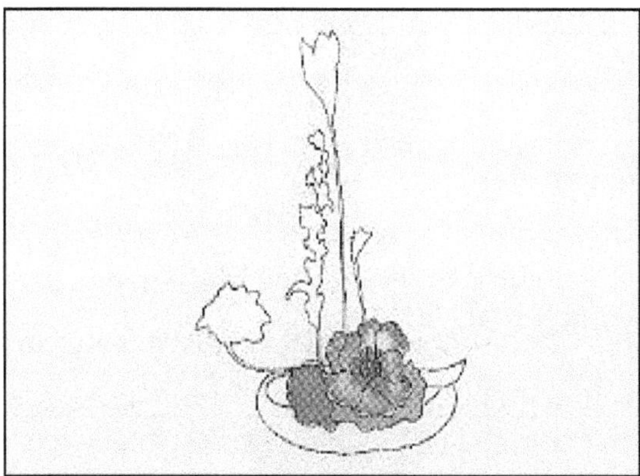

3. Fill in with filler flowers as needed.

g. The Lazy "S" or "Hogarth's Curve"

1. Anchor the foam securely. Bend the stems gently into graceful curves and insert them in place so they balance.

2. Add the focal flowers following the lines of the upper and lower curves.

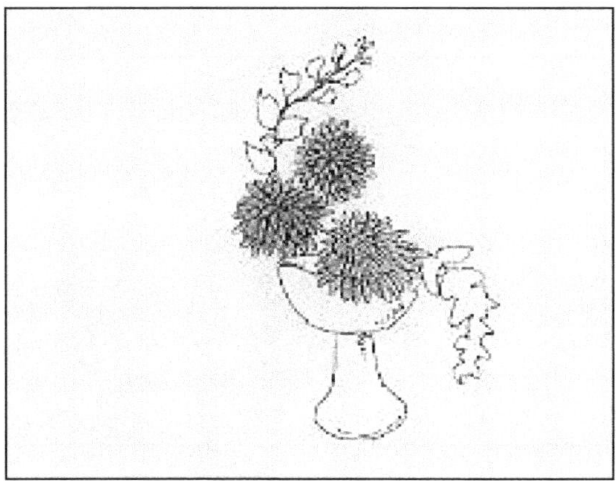

3. Cluster filler blossoms and foliage around the central flowers maintaining the rhythm of the 'S'.

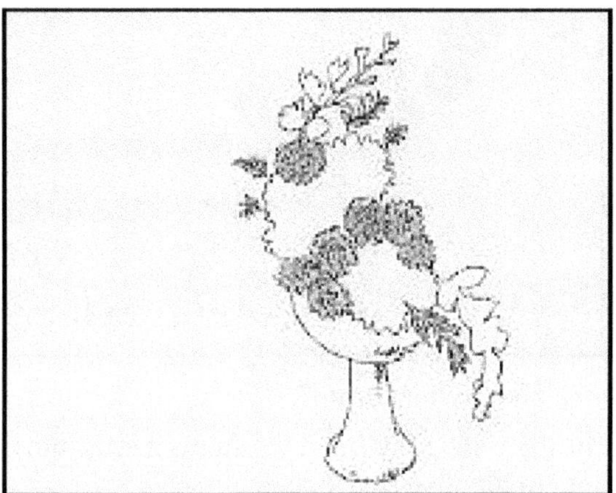

h. Free Standing Arrangements

1. Fill a shallow container with foam that extends one inch over the top. Secure the foam with hot glue, floral clay or floral tape. Define the shape of the design with the line flowers and leaves.

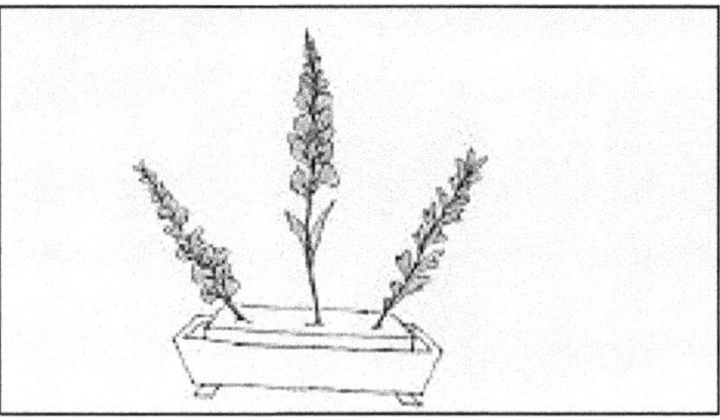

2. Place the focal flowers and leaves, turning the vase as you go so all the sides are even.

3. Add filler flowers to integrate the design.

4.2 Containers and Requirements for Flower Arrangements, Free Style, Shallow and Mass Arrangement

Basic Designs in making Flower Arrangements

There are three basic designs for flower arrangements *viz.*, Line Arrangements, Mass Arrangements and Line-Mass Arrangements.

a. Line Arrangement of Flowers

Designs composed primarily of line material have minimum flowers and foliage established in the focal area. The open form emphasizes the beauty of the plant material. A bare branch, a piece of decorative wood or a pine branch creates a well-defined line design. The addition of a few flowers and leaves is all that is needed to achieve a dramatic effect.

b. Mass Arrangement of Flowers

Full, flowing and symmetrical, a generous mixture of flowers arranged in a decorated China vase portrays the typical mass design so reminiscent of Victorian opulence and elegance. Mass designs are usually triangular, oval, and circular or fan shaped.

c. Line-Mass Arrangement of Flowers

Line-Mass creations employ the use of mass for a greater feeling of depth. Plant material is placed to form an orderly outline and massed to create a focal area with gradual changes from line at the periphery of the arrangement to mass within the central axis. Line-mass designs have an open form with symmetrical or asymmetrical balance.

Other Arrangements

Vertical Designs

These are easily assembled in tall, narrow containers. The dominant thrust of the plant material is vertical with few lines leading in other directions. Cattails, delphinium, anthurium, gladiolus, iris or ting-ting are naturally tall and straight. Focal material is placed at the lip and slightly above within the container. Vertical designs create feelings that are uplifting and aspiring.

Horizontal Arrangements

These are often used for centerpieces, long tables, windowsills, anywhere an extended arrangement is suitable. Use a shallow container and position the sprays of line flowers to establish the length of the design. Insert the focal flowers in the middle so they droop over the lip of the container on both sides and reach towards the line material. Fill in with filler flowers around the focal area.

Taller Horizontal Arrangements

These make interesting viewing when the flowers and foliage are stacked by height from back to front. Use a rectangular basket or vase. Insert the line flowers along the back of the container, the filler floral in the middle and the focal flowers at the bottom. Horizontal arrangements can be very playful and inviting.

Triangular Arrangements

These may be equally balanced on each side or asymmetrical with one point of the triangle extending further than the other. The stems radiate from a central area with paler and smaller flowers and leaves extending to the outer edges. Deep colored or brighter blooms used near the center or slightly below give a feeling of stability and provide a focal point of interest. Make the height higher than the width, and position the flowers so they droop on the periphery in an elegant and relaxed fashion. Fill in with small flowers and foliage keeping within the triangular shape.

The Asymmetrical Triangle

This is a very popular line or line-mass design. The design may be either right or left handed, meaning the tallest line will either be right or left of the axis. Some bare branches, a few flowers and some leaves will suffice to make a beautiful design.

Oval Arrangements

These are designed to look good from all sides. They can be constructed with larger blooms centrally placed and smaller ones echoing around the edges. Tape your vase with floral tape to a lazy Susan. It will be easier to construct if you can turn it as you add plant material.

An Oblique Design

Its depends upon dynamic balance. The main line is on an incline, slanting somewhere between the perpendicular and the horizontal. Stabilize the design by placing the visually heavier material near the axis. Short footed or flat containers provide a good base for the strong thrust of an oblique design.

The Zigzag Design

It is restless and quick as lightning. A small amount of plant material is used to emphasize the focal area, where the line originates, and may extend along the line trailing off as it leads the eye toward the sharp angles.

Minimal Arrangements

Its make use of space as a design element. Oriental art is rich and evolved in the use of space and precise placement of plant material.

The "S" Curve

Or what William Hogarth called **"the line of beauty"** in his paintings, is a very popular style. The design may be constructed upright or horizontally. The graceful sweep of the "S" is usually elongated. Select line flowers or foliage that has graceful curves. Insert in place so they balance. Add the focal flowers following the lines of the upper and lower curves. Cluster filler flowers around the central flowers.

Free Standing Arrangements

These are similar to ovals, but ovals have a gently rounded arc at the top, and free standing arrangements come to a point like a Christmas tree. They include an abundance of flowers.

4.3 Japanese Type of Arrangement: Ikebana (Japanese/Oriental Flower Arrangement

The word literally means '*making flowerslive*' in Japanese. Ikebana is the ancient Japanese art of flower arranging. The name comes from the Japanese *ike*, meaning 'alive' or 'arrange' and *bana* meaning 'flower.' This Japanese style has been practiced for thousands of years. These arrangements are more than an aesthetic grouping of plant materials. They are symbolic representations of an ideal harmony that exists between earthly and eternal life. In each arrangement, there is an imaginary triangle. Its tallest line represents 'heaven'. Facing and looking towards heaven is 'man'. The lowest line, looking up to both, is 'earth'. Ikebana, the Japanese word for the art of flower arrangement, comes from two Japanese words, *IKE*, meaning pond and *HANA*, meaning flower, the pond flower. Thus, Ikebana is the art of arranging flowers in water. Ikebana emphasizes most on measurement, the scale of floral arrangement. The main aim of Ikebana is to bring nature indoors and make it live in the floral arrangement that is being made.

Originally, the first school of ikebana was divided into three styles.

- ✰ The Shin - The formal style involved an erect linear arrangement in a bronze container, arranged on a carved teakwood stand.

- ☆ The Gyo, or semi-formal school - The school features flowing, sweeping lines and the use of a variety of containers.
- ☆ The So, or informal school - The school displayed flowers in bamboo vases, baskets, and natural wood containers as well as in pottery. The flowers are seated on bases of bamboo or natural wood.

Principles of Ikebana

Ikebana has become an art form that is associated with a meditative quality. Creating an arrangement is supposed to be done in silence to allow the designer to observe and meditate on the beauty of nature and gain inner peace. Seasoned designers realize not only the importance of silence, but also the importance of space, which is not meant to be filled, but created and preserved through the arrangements. This ties into other principles of Ikebana including minimalism, shape and line, form, humanity, aesthetics, and balance.

4.4 Bouquet and Garland Making

1. Bouquet

A flower bouquet is a collection of flowers in a creative arrangement. Flower bouquets can be arranged for the decor of homes or public buildings, or may be handheld. Handheld bouquets are classified by several different popular shapes and styles, including nosegay, crescent, and cascading bouquets. Flower bouquets are often given for special occasions such as birthdays or anniversaries. They are also used extensively in weddings. Bouquets arranged in vases or planters for home decor can be arranged in either traditional or modern styles. Symbolism may be attached to the types of flowers used, according to the culture

Materials Required

30 to 60 stems of a hardy flower like the rose (20 to 40 for each bridesmaid bouquet). Bucket, Paper towels,Ribbon (in a complementary color), 1 to 2 inches wide,Rubber bands or green waxed floral tape,Stem cutter or very sharp knife,Stem stripper, Straight pins or pearl-tipped corsage pins.

Step 1: Preparing the Flowers

Use your hands or a stem stripper to remove excess foliage and thorns, and pull off damaged or unattractive outer petals. Fill a sink or bucket with water, and holding the stems underwater use the stem cutter or knife to cut the stems at an angle about 2 inches from the bottom. Allow the flowers to drink for a few seconds with the stem ends underwater, then place the stems in a bucket filled halfway with cool water until you are ready to use them.

Step 2: Assembling the Flowers

Take one stem at a time with one hand, and use your other hand to hold the flowers in place. Assemble four flowers at an even height in a square shape — these will be at the center of the dome. Arrange the other flowers one by one around the center flowers to create a dome shape. See other wedding bouquet shapes.

Step 3: Securing the Bouquet

Use a rubber band or floral tape to bind the stems at the spot where they naturally join (about 3 to 4 inches below the flower heads). Repeat the binding toward the end of the stems, leaving about 2 inches of excess stem beneath the bind to trim later. Either place the stems in water and wrap them later or continue to Step 4.

Step 4: Finish the Handle

Cut the stem ends so they are all the same length, about 7 to 8 inches long. Dry off the stems with a paper towel. Cut a length of ribbon about three times as long as the length of the stems. Tuck the end of the ribbon inside the top bind and start wrapping in a spiral down the length of the stem. When you reach the bottom, wrap in a spiral back up the stem. At the top, tuck the cut end of the ribbon underneath and secure with a couple of pins pushed through the ribbon and into the stems.

Step 5: Preserving the Bouquet

Wrap the bouquet in tissue.Store it in the refrigerator until you are ready to leave for the ceremony.

2. Garlands

Garlands are a perfect and highly versatile addition to your wedding decor. Use them on your wedding arch, drape them over chairs, hang them on tables or over doors and in car. This gallery will give show you how to use the same flower to create 3 different garlands, each with a unique look. We have used a white chrysanthemum for our main flower, however these garlands can be adapted for any kind of flower. Garlands used as floral decoration by women for their hair, or as religious symbols to honor gods, or during auspicious celebrations, or even as a symbol of respect to a guest.

How to Make Floral Garlands

Basic Carnation Garland

To honor a guest, give to a loved one, decorate for an event or make for yourself. You will need about 4 dozen carnation heads and a needle and thread.

1. Cut the flower heads at the base of the calyx.

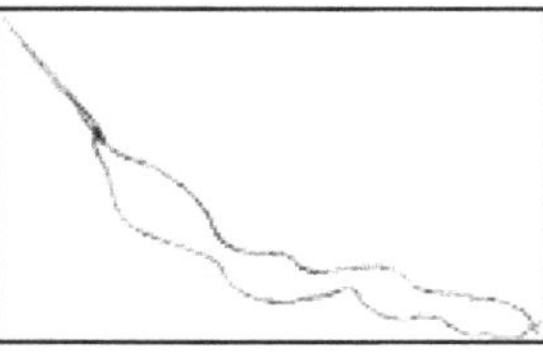

2. Measure a double length of thread the size of the garland plus 10" for knotting. Thread your needle.

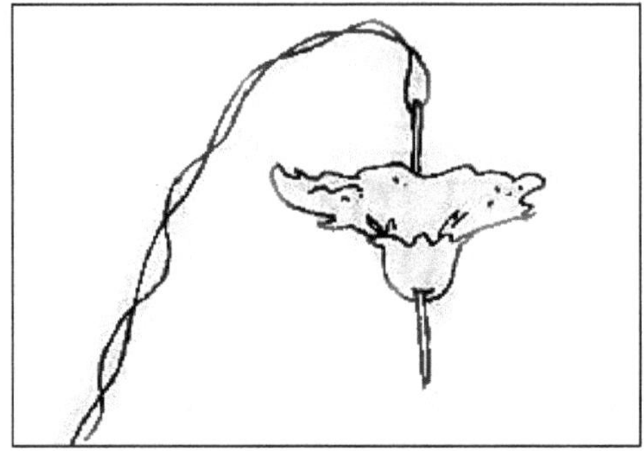

3. Divide the flower heads in half. Insert the needle through the flower from the head to the stem. Use half the flowers.

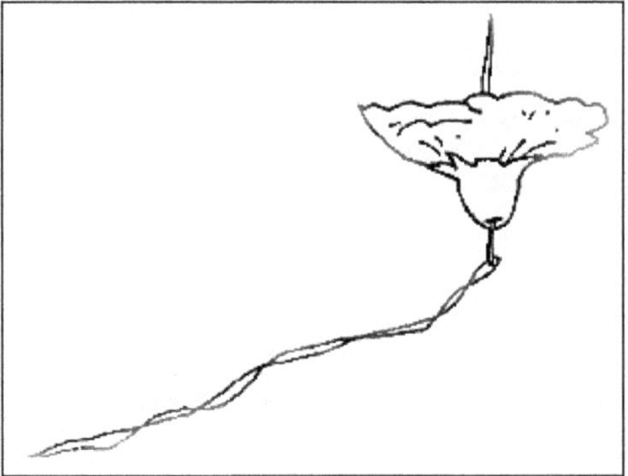

4. Insert the needle through the stem end up through the flower head using the remaining flowers.

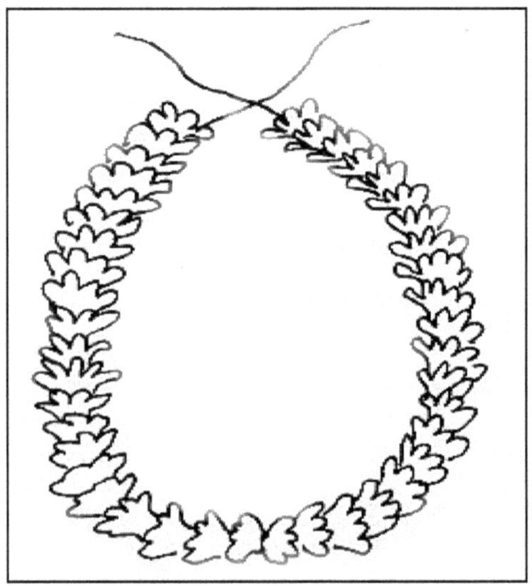

5. Compress the flowers on the thread before tying the garland into a circle with four knots.

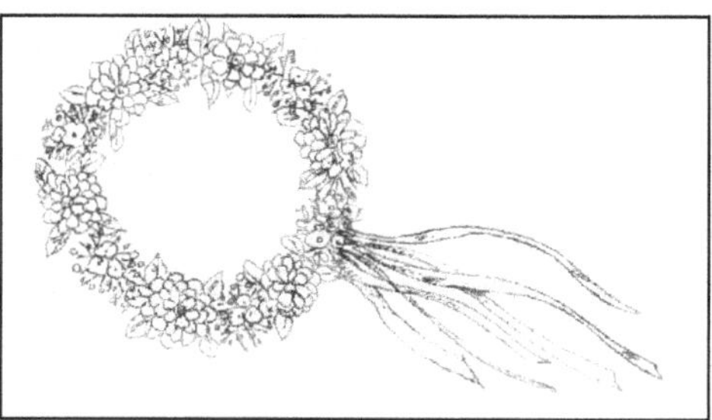

4.5 Dry Flower Arrangement

Dried flower arrangements have been popular for many centuries. Well-liked all through the year, they are more common during winter months for the obvious reason that in those seasons fresh flowers are not available. Flowers can be dried using various methods such as air drying, water drying, oven drying, drying in sand, drying with desiccants, and microwave drying.

A Simple and Beautiful Dried Flower Arrangement

Fix your vase using floral clay on a revolving tray that is commonly used for placing foods at the center of a dining table (*lazy susan*). This is to help you check out the flower arrangements from all sides.

- ✰ Force a piece of foam into the vase. Line material such as lavendar and straw may be used in order to fix the total appearance.
- ✰ Insert the bigger focal flowers like hydrangea first, followed by average-sized flowers like dried roses. Space the flowers uniformly all through.
- ✰ Add supplementary line material to give an emphasis to the overall shape. Fill up vacant spaces using fern and add flax for creating an appearance of completeness.

How to make a Dried Flower Arrangement

The basic materials you need for making a wreath are straw and Styrofoam, wired wooden floral picks, pliers, white glue, greening pins, and a 21 gauge wire (that is, a wire of diameter 0.0285 inches).Now your task is to join the leaves, cones, and other decorations to the base.

Step1

Join a ring of 0.028 inch diameter wire to the wreath foundation.

Step 2

Pull out the plastic wrap from the wreath foundation so that the floral picks and greening pins can go in easily.

Step 3

With the wreath foundation as the base, start decorating from its outside edge and work toward the center. This will result in a fuller and a more natural appearance.

Step 4

Test for completeness by keepin the wreath on a table and verifying that there is no space between the decorating material and the surface of the table.

4.6 Harvesting Methods and Storage

i. Harvesting

The process of cutting and gathering a crop is called harvesting. The traditional implement used to harvest a crop is the sickle. Modern farms use a harvester, which cuts the crop. A harvester can be combined with other machinery that threshes and cleans the grain as well. It is then called a combine harvester or combine. Harvesting requires art and practice because a large proportion of crop can be lost due to an improper method of harvesting.

Steps of Harvesting of Fruits and Vegetables

i) Identification and judging the maturity of fruits.
ii) Selection of mature fruits.
iii) Detaching or separating of the fruits from tree, and
iv) Collection of matured fruits.

Method of Harvesting

Different kinds of fruit and vegetables require different methods after harvesting. The methods of harvesting are:

1. Manual Harvesting
2. Mechanical Harvesting

1. Manual Harvesting

Harvesting by one's own hand is called manual harvesting. It is done in several ways:

a. Ladder/bag picking method
b. Poles/Clippers method
c. Harvesting by means of cutting knives
d. Harvesting by means of digging tools.

2. Mechanical Harvesting

In this method numbers of mechanical devices are used for harvesting the produce on commercial scale.

ii. Storage

In the case of small-scale cultivation, farmers use the harvested crop for themselves while large-scale production is mainly for marketing. Thus the cultivators have to store the grains. For this, a proper storage space has to be arranged. Inadequate storage space and improper storage methods can lead to a huge grain loss.

Storage Methods

a) Traditional Drying/Storage Systems

Many farmers continue to store their produce in the drying place. Often the root or the eaves are still full with maize even after the produce has dried.

b) Drying Cribs

Many agricultural books say that the drying crib can also be used for a storage barn. However, as we have seen in the above section, it is too dangerous to leave the grain exposed to insects, birds, and other pests. After the grain is dry it should be moved to a better storage place.

c) Bag Storage

This is a very popular form of storage. Transportation of the grain is done in the same jute bag, the bags are easy to handle and the jute bag allows you to store different grains in the same room.

d) Bamboo Boxes

The box is constructed entirely of raffia bamboo sticks and bamboo rope. The floor of the box is raised off the ground so that water cannot rise up from the ground and enter the box. The bottom of the walls are often packed with mud soil to discourage rats. The box has either a zinc/grass roof or is placed under the eaves of the house to keep the rain off the box.

e) Drums (Air Tight Storage)

A very good, but more expensive method is to use old oil drums. The drums should be well cleaned. All holes should be repaired and sealed properly with sodden.

f) Others

Baskets, tins, and empty calabashes can also be used to store grain. Just ensure that the grains and the containers are clean and free of insects. Keep the container in a clean, cool, dry place. Baskets, tins' end calabashes are small and are ideal for seed storage. However, for large amounts of grain bigger containers are needed.

4.7 Marketing of Fruits, Vegetables and Flowers

Marketing is the processing of selling products. If you have a big farm or live far away from customers, you might want to sell your crops to one place, such as a store, restaurant, and food co-op. This is called wholesale marketing. Selling your crops this way is faster than other ways of selling. It is also a good choice if you are not comfortable meeting and talking to a lot of people. Wholesale marketing will probably not make you as much money as selling directly to the customer, and you may not get paid right away. But it is a simple, easy way of selling.

Types of Market

a. The Farmer's Market

These are the market maintained by farmers. Many people come to the farmer's market to buy vegetables. You can sell a lot of vegetables, so you need to bring enough vegetables.

b. Roadside Market

People like to stop at roadside stands to buy fresh vegetables. Roadside stands can be close to your home or garden. You can sell your crops without traveling far, and you can make money selling only one or two crops. For instance, fresh-picked sweet corn, tomatoes, and pumpkins sell well.

c. Pick your Own or You-Pick

Some customers like going into fields and picking their own vegetables. This is sometimes known as a You-Pick or "U-Pick" field. The best customers for a you-pick field need a lot of vegetables for canning or freezing. If you are near people who can or freeze vegetables, selling this way may be a good choice.

4.8 Preservation and Process: Methods of Reducing Deterioration Sing of Fruits and Vegetables

Different Methods Storage

Physical

- ✰ Heating
- ✰ Cooling
- ✰ Lowering of water content Drying/dehydration. Concentration
- ✰ Sterilising filtration
- ✰ Irradiation
- ✰ Other physical means (high pressure, vacuum, inert gases)

Chemical

- ✰ Salting
- ✰ Smoking
- ✰ Sugar addition
- ✰ Artificial acidification

- ☆ Ethyl alcohol addition
- ☆ Antiseptic substance action

Biochemical

- ☆ Lactic fermentation (natural acidification)
- ☆ Alcoholic fermentation

Procedures for Fruit and Vegetable Preservation

Procedures	*Practical Applications*
Fresh storage	Fruits, vegetables
Cold storage	Fruits, vegetables
Freezing	Fruits, vegetables
Drying/dehydration	Fruits, vegetables
Concentration	Fruit and vegetable juices
Chemical preservation	Fruit semi-processed
Preservation with sugar	Fruit products/preserves
Pasteurization	Fruit and vegetable juices
Sterilisation	Fruits, vegetables
Sterilising filtration	Fruit juices
Irradiation	Fruits, vegetables

Types of Storage/Preservation

a. Preservation with Sugar

The principle of this technology is to add sugar in a quantity that is necessary to augment the osmotic pressure of the product's liquid phase at a level which will prevent microorganism development. From a practical point of view, however, it is usual to partially remove water (by boiling) from the product to be preserved, with the objective of obtaining a higher sugar concentration. In concentrations of 60 per cent in the finished products, the sugar generally assures food preservation. It is important to know the ratio between the total sugar quantity in the finished product and the total sugar concentration in the liquid phase because this determines, in practice, the sugar preserving action.

b. Heat Preservation/Heat Processing

There are various degrees of preservation by heating; a few terms have to be identified and understood.

1. *Sterilisation*: By sterilisation we mean complete destruction of micro-organisms. Because of the resistance of certain bacterial spores to heat, this frequently means a treatment of at least 121°C (250°F) of wet heat for 15 minutes or its equivalent. It also means that every particle of the food must receive this heat treatment. If a can of food is to be sterilised, then immersing it into a 121°C pressure cooker or retort for the 15 minutes will not be sufficient because of relatively slow rate of heat transfer through the food in the can to the most distant point.
2. "*Commercially Sterile*": Term describes the condition that exists in most of canned or bottled products manufactured under Good Manufacturing Practices procedures and methods; these products generally have a shelf-life of two years or more.
3. Pasteurized means a comparatively low order of heat treatment, generally at a temperature below the boiling point of water. The more general objective of pasteurization is to extend product shelf-life from a microbial and enzymatic point of view; this is the objective when fruit or vegetable juices and certain other foods are pasteurized. Pasteurization is frequently combined with another means of preservation - concentration, chemical, acidification, *etc.*
4. Blanching is a type of pasteurization usually applied to vegetables mainly to inactivate natural food enzymes. Depending on its severity, blanching will also destroy some In a simplified manner, the main operations employed in heat preservation can be described as follows:

c. Pasteurization

a) Low pasteurization where pasteurization time is in the order of minutes and related to the temperature used; two typical temperature/time combinations are as following:

63°C to 65°C over 30 minutes or

75°C over 8 to 10 minutes.

Pasteurization temperature and time will vary according to:

- Nature of product; initial degree of contamination;
- Pasteurized product storage conditions and shelf life required.

In this first category of pasteurization processes it is possible to define three phases:

- Heating to a fixed temperature;
- Maintaining this temperature over the established time period (= pasteurization time);
- Cooling the pasteurized products: natural (slow) or forced cooling.

b) Rapid, high or flash pasteurization is characterized by a pasteurization time in the order of seconds and temperatures of about 85° to 90°C or more, depending on holding time. Typical temperature/time combinations are as follows:

88°C (190°F) for 1 minute

100°C for 12 seconds

121°C for 2 seconds

d. Food Irradiation

Food irradiation is one of the food processing technologies available to the food industry to control organisms that cause food-borne diseases and to reduce food losses due to spoilage and deterioration. Food irradiation technology offers some advantages over conventional processes. Each application should be evaluated on its own merit as to whether irradiation provides a technical and economical solution that is better than traditional processing methods.

This classification of methods of reducing deterioration presents some difficulties because their preservation effects are physical, physico-chemical, chemical and biochemical complex phenomena which rarely act in isolation. Normally they take place together or one after the other.

e. Freezing

Many vegetables keep well in the freezer. When blanched and frozen soon after harvesting, this can be the best method for retaining nutrients, as well as color, texture and flavor. Most vegetables can last 8 - 12 weeks in the freezer. Cold storage can be combined with storage in an environment with added of carbon dioxide, sulphur dioxide, *etc.* according to the nature of product to be preserved.

☆ Cold storage of dried/dehydrated vegetables in order to maintain vitamin C; storage temperature can be varied with storage time and can be at −8°C for a storage time of more than one year, with a relative humidity of 70-75 per cent.

☆ Packaging under vacuum or in inert gases in order to avoid action of atmospheric oxygen;, mainly for products containing beta-carotene.

☆ *Chemical preservation*: a process used intensively for prunes and which has commercial applications is to rehydrate the dried product up to 35 per cent using a bath containing hot 2 per cent potassium sorbate solution. Another possible application of this combined procedure is the initial dehydration up to 35 per cent moisture followed by immersion in same bath as explained above; this has the advantage of reducing drying time and producing minimum qualitative degradation. Both applications suppress the dehydrated products reconstitution (rehydration) step before consumption.

f. Canning

Canning is a great method for preserving fruits and vegetables with a high water content, like tomatoes, mushrooms, beans and peaches, but it is essential you follow canning guidelines to the letter.

g. Dehydrating (Drying)

Drying fruits, vegetables and herbs is also a very easy process and can be done without any special equipment or speeded up by using the oven or a dehydrator. Dehydrating removes all the water from a food and because it lacks moisture, mold and bacteria can't grow on it. Dehydrated foods will last about four months to a year, but some nutrients will be lost in the process. Commonly dried foods include meats, fruits (either in their original form or pureed to make fruit leathers or bars), herbs and seeds. Dried fruits, seeds, jerky, leather and even popcorn can be done by these methods.

h. Pickling

Pickling, which uses salt and/or vinegar to inhibit the growth of bacteria, is one of the oldest methods of food preservation. While most of us think of sweet pickled cucumbers, sauerkraut, relishes and fruits can also be pickled. Pickled foods will last anywhere from 3 months to a year.

i. Jams and Jellies

In hot, arid regions, sun drying is an option, but it demands at least 3-4 sunny days of 100-degree heat in a row. The easiest and most effective way of drying your foods is to use a commercially made dehydrator. These have several levels of stacking trays that allow air to circulate in and around the foods at just the right temperature–high enough to dehydrate the food but low enough not cook it. Generally, the foods are laid out on the trays and, according to the manufacturer's instructions, you'll set the time, temperature and position of the trays. Dehydrators can take several hours or days to dry foods completely. Once dried, keep all foods tightly sealed in a container in a cool, dark place to ensure its longevity. Some food preservation books and "raw" food cookbooks also include detailed instructions for using a conventional oven, set at a low temperature with the door cracked, as a food dehydrator, which is a great option if you're not ready to invest in your own dehydrator.

Chapter 5

GROWTH REGULATION AND PLANT PROTECTION

5.1 Growth Regulators in Horticulture

Plant growth regulators (also called plant hormones) are numerous chemical substances that profoundly influence the growth and differentiation of plant cells, tissues and organs. Plant growth regulators may be defined as any organic compounds, which are active at low concentrations (1-10 ng/nl) in promoting, inhibiting or modifying growth and development. Plant growth regulators function as chemical messengers for intercellular communication. There are currently five recognized groups of plant hormones they are auxins, gibberellins, cytokinins, abscisic acid (ABA) and ethylene. They work together coordinating the growth and development of cells. Ethylene is mainly involved in abscission and flower secscence in plants and is rarely used in plant tissue culture. In addition to the five principal growth regulators, two other groups sometimes appear to be active in regulating plant growth, the brassinosteroids and polyamines.

Uses of Growth Regulators in Horticultural Crops

1. Propagation

They are applied in the form of paste and solution. The concentration of the chemical varies with plant species and types of cutting and method of application.

2. Seed Germination

GA significantly accelerates seed germination in many plant species. Pre soaking the seed with G. Such as bhendi and sugar beet increase germination.

3. Induction of Flowering

Plant growth regulators like NAA at 10 to 50 ppm causes early flowering in pine apple. 2,4-D at 6 to 10 ppm has used to induce flowering in pine apple. Flowering can be delayed by 1 to 2 weeks NAA at 200 to 800 ppm application in apple, cherries, pears, peaches, and plums.

4. Sex Expression

Plant growth regulators can change the sex of the flowers. Male sterility can be induced in corn by MH 9 malic hydrozide). It is used in plant breeding for induction of male sterility. Application of NAA, IAA and GA at 50 to 100 ppm increases female flowers in pumpkin, cucumber to get more yield.

5. Flower and Fruit Thinning

Many fruit trees produces heavy flowering and fruit in one year and few or one in next year. By using G.R the normal bearing can be maintained NAA at 5 to 10 ppm and NAA at 5 to 7 ppm for thinning of apple, peaches and grapes.

6. Pre Harvest Drop of Fruits

Flower and fruit drop is a problem in many fruit crops. Application of NAA 10- 50 ppm in mango, citrus and chilies reduce fruit drop by preventing formation of abscission layer.

7. Fruit Development

Application of 50- 100 ppm GA in grapes increases the berry size.

8. Early Maturity

Early maturity fetches higher prices in the market. In pine apple application of 20 ppm NAA induces early flowering and early maturing at least by two months. Spraying of 50 ppm NAA reduces maturity in grapes, use of 250 – 400 ppm of Ethrel induces early maturity in Ber.

9. Early Ripening and Colour Development

Fruits like mango, banana, papaya ripes after harvest. Dipping of fruits in 20-50 ppm Ethrel solution induces golden yellow colour to fruit induces early maturity.

10. Delayed Maturity

Delay in ripening is required when fruit are to be sent to long distance market. Dipping of fruit in 2,4-D, 2,4- 5- T or MH- 40 extends storage life of fruits.

11. Sprouting of Bud

Ethrel, GA, thio urea, IBA and Cyotkininn, spray induces sprouting of buds.

12. Braking of Dormancy

GA, Ethrel, NA are used in breaking dormancy in seeds and buds. Some Defination:

Classification of Growth Regulators

1. **Auxins**
2. **Gibberellins**
3. **Cytokinenins**
4. **Etylene**
5. **Dormins** (Abscissic Acid (ABA), Phaseic Acid)
6. **Flowering Hormones** (Florigin, Anthesin, Vernalin)
7. **Phenolic Substances** (Coumarin)
8. **Miscellaneous Natural Substances** (Vitamins, Phytochrome Tranmatic Substances(
9. **Synthetic Growth Retardants** (Ccc, Amo, 1618, Phosphin - D, Morphacting, Malformis)
10. **Miscellaneous Synthetic Substances** (Synthetic Auxins, Synthetic Cytokinin)

Auxins

Auxins stimulate cell elongation and influence a host of other developmental responses, such as root initiation, vascular differentiation, tropic responses, apical dominance and the development of auxiliary buds, flowers and fruits. Auxins are synthesized in the stem and root apices and transported through the plant axis. The principal auxin in plants is indole-3-acetic acid (IAA). Several other indole derivatives, all as precursors to IAA, are known to express auxin activity, probably by converting to IAA in the tissue. Auxins in plant tissue culture are used to induce callus from explants, and cause root and shoot morphogenesis. Auxins are often most effective in eliciting their effects when combined with cytokinins.

Cytokinins

Cytokinins are able to stimulate cell division and induce shoot bud formation in tissue culture. They usually act as antagonists to auxins. (Cytokinins are N6 substituted derivatives of the nitrogenous purine base adenine.) Cytokinins most used in tissue culture include zeatin, adenine, 6-(g,g-dimethylallylamino)purine (2 iP) and kinetin. Cytoknins often inhibit embryogenesis and root induction.

Gibberellins

The main effect of gibberellins in plants is to cause stem elongation and flowering. They are also prominently involved in mobilization of endosperm reserves during early embryo growth and seed germination. Gibberellins are an extensive chemical family based on the ent-gibberellane structure. There exit over 80 different gibberellin compounds in plants but only giberrellic acid (GA3) and GA4+7 are often used in plant tissue culture. In tissue culture, gibberellins are used to induce organogenesis, particularly adventitious root formation.

Abscisic Acid

Abscisic acid (ABA) in plants is a terpenoid involved primarily in regulating seed germination, inducing storage protein synthesis and modulating water stress. In plant tissue culture, it is used to help somatic embryogenesis, particularly during maturation and germination.

Ethylene

Ethylene is a simple gaseous hydrocarbon with the chemical structure $H_2C=CH_2$. Ethylene is apparently not required for normal vegetative growth. However, it can have a significant impact on development of root and shoots. Usually, ethylene is not used in plant tissue culture.

Plant Growth Regulator

Class	*Function(s)*	*Practical Uses*
Auxins	Shoot elongation	Thin tree fruit, increase rooting and flower formation
Gibberellins	Stimulate cell division and elongation	Increase stalk length, increase flower and fruit size
Cytokinins	Stimulate cell division	Prolong storage life of flowers and vegetables and stimulate bud initiation and root growth
Ethylene generators	Ripening	Induce uniform ripening in fruit and vegetables
Growth inhibitors	Stops growth	Promote flower production by shortening internodes
Growth retardants	Slows growth	Retard tobacco sucker growth

5.2 Rooting Hormone, Growth Promoters, Flower Induction and Parthenocarpy

1. Rooting Hormone

A rooting hormone is a naturally occurring or synthetic hormone that stimulates root growth in plants. Auxins are one of the main plant hormones that aid in the creation of initial root growth. More specifically, indole acetic acid (IAA) is the natural auxin found in plants that is responsible for natural root stimulation. IAA is involved in just about every aspect of plant growth and development, including the formation of embryo development, induction of cell division, stem elongation, vascular tissue differentiation, fruit/flower development, tropic behaviors (leaves and stems moving toward the light source) and the induction of rooting.

Types of Rooting Hormones

a. Liquid

Liquid is by far the most common type of rooting hormone, but there are two different formats it's sold in. The first is standard-strength rooting hormone that can be used right out of the bottle.

b. Powder

When using powder rooting hormone, dip your cuttings into water first so the powder will adhere to and seal the cut area. Then, pour some powder into a separate bowl or plate so as to avoid contamination. Finally, dip your wet cuttings into the powder and shake off any excess.

c. Gel

The most popular of the three types of rooting hormone is the gel form.

2. Plant Growth Promoter

These substances are formed in one tissue or organ of the plant and are then transported to other sites where they produce specific effects on growth and development. They are referred to as plant hormones. Plant hormones are organic compounds which are capable of promotion, inhibition or modification of growth. The plant hormones are also known as growth factors, growth hormones, growth substances, growth regulators or phytohormones. Phytohormones are grouped into 2 main types - growth promoters have a positive effect on a process and thus promote it, whereas the growth inhibitors have a negative effect and cause inhibition.

a. Auxin

Auxins are the best known plant growth regulators. They promote growth of stem or coleoptile sections and decapitated (apex removed) coleoptiles, but in the same concentration are incapable of causing growth in intact plants.

Types of Auxins

There are two major types of auxins:

(i) Natural auxins and

(ii) Synthetic auxins

(i) Natural Auxins

These are naturally occurring auxins in plant parts and, therefore, regarded as phytohormones. The best known and universally present natural auxin is Indole-3-acetic acid. Other natural occurring auxins are hidole-3-pyruvic acid, Indole-3-ethanol, Indole-3-acetaldehyde, *etc.* The natural auxins may occur in plants in the form of free auxins and bound auxins.

(ii) Synthetic Auxins

These are the chemicals synthesized by chemists that cause various physiological actions similar to IAA. They are not treated as phytohormones but are considered as plant growth regulators. Some of the synthetic auxins are: Indole-3-butyric acid (IBA), Indole-3- propionic acid (IPA), α- and β-naphthalene acetic acid (NAA), 2, 4- dichlorophenoxy acetic acid (2, 4-D), benzoic acids, *etc.*

AUXIN	USES
NAA (Naphthalene acetic acid)	(*i*) Induces rooting of cuttings (*ii*) Is used for preventing sprouting of potatoes. (*iii*) Foliar spray causes flowering in litchi and pineapple. (*iv*) Helps natural fruit setting.
IPA (Indole propionic acid)	(*i*) Induces rooting of cuttings. (*ii*) Helps natural fruit setting.
IBA (Indole butyric acid)	(*i*) Induces rooting of cuttings.
2, 4-D (2, 4-dichlorophenoxy-acetic acid)	(*i*) Selective weed-killer — kills broad-leaved plants (dicots) but does not affect mature monocots. (*ii*) Causes flowering in litchi and pineapple. (*iii*) Prevents fruit drop (abscission)
2, 3, 6-trichlorobenzoic acid	Powerful weed-killer.
2, 4, 6-trichlorobenzoic acid	—Do—

b. Gibberellins

Gibberellins, like auxins, are chemicals which have the capacity to influence both growth and development of plants. The Japanese consider tall people foolish and hence they called those tall seedlings of rice 'bakanae' (foolish seedlings). Later it was found that it was not a new variety but the tallness was caused due to the infection of a fungus *Gibberella fujikuroi.*

c. Cytokinins

A third type of plant hormone, the cytokinins stimulate growth of cells in tissue culture or organ culture and have a marked increasing effect on cell division. The existence of specific substances which can control cell division in plants was suspected many years before such substances were finally discovered. Synthetic Kinetin was found to be a very potent promoter of cell division in tobacco pith. Later on it was discovered that substances division promoting activity were widely distributed in plants. Letham isolated it from immature seeds of maize and named it Zeatin.

3. What is Parthenocarpy?

In botany and horticulture, parthenocarpy (literally meaning virgin fruit) is the natural or artificially induced production of fruit without fertilization of ovules. Parthenocarpy (or stenospermocarpy) occasionally occurs as a mutation in nature; if it affects every flower the plant can no longer sexually reproduce but might be able to propagate by apomixis or by vegetative means. Parthenocarpy, development of fruit without fertilization. The fruit resembles a normally produced fruit but is seedless. Varieties of the pineapple, banana, cucumber, grape, orange, grapefruit, persimmon, and breadfruit exemplify naturally occurring parthenocarpy. Seedless parthenocarpic fruit can be induced in nonparthenocarpic varieties and in naturally parthenocarpic varieties out of season by a type of artificial pollination with dead or altered pollen or by pollen from a different type of plant. The application of synthetic growth substances in paste form, by injection, or by spraying, also causes parthenocarpic development.

Types of Parthenocarpy

Parthenocarpy can be categorized into two parts, which are

1. Vegetative Parthenocarpy

This type of parthenocarpy generally takes place without pollination and due to the absence of pollination, no seeds are formed.

2. Stimulative Parthenocarpy

This type of parthenocarpy takes place without the process of fertilization. This condition occurs when the ovipositor of a wasp is inserted into the ovary of a flower. This can also be achieved by flowing air or growth regulators into the unisexual flowers that are present inside the syconium.

5.3 Plant Protection: Common Diseases of Fruit and Vegetables (Mango, Tomato)

Plant Protection continues to play a significant role in achieving targets of crops production. Plant Protection The major thrust areas of plant protection are promotion of Integrated Pest management, ensuring availability of safe and quality pesticides for sustaining crop production from the ravages of pests and diseases, streamlining the quarantine measures for accelerating the introduction of new high yielding crop varieties, besides eliminating the chances of entry of exotic pests and for human resource development including empowerment of women in plant protection skills.

Plant protection is the science and practice of managing pests, diseases and weeds that damage crops and other plants, and which can have a devastating effect on farmer livelihoods.

Following are the direct forms of plant protection:

1. Physical methods
2. Chemical methods
3. Biotechnical methods
4. Biological methods
5. Integrated methods

i. Major Diseases of Vegetables

a. Club Root

A serious problem in home gardens, club root infects brassica crops – cabbage, broccoli, cauliflower, *etc.* – which wilt during the heat of the day. Older leaves turn yellow and drop. Roots are distorted and swollen.

b. Downy Mildew

Affects many vegetables and appears as a white to purple "downy" growth on the undersides of leaves and along stems. The best way to prevent downy mildew is to avoid the conditions that favor it.

c. Early Blight

Symptoms of early blight include brown and black spots on leaves that enlarge and develop rings like a target. Leaves may actually die. a long period of wet weather.

d. Late Blight

True to its name, this fungal disease occurs later in the growing season with symptoms often appearing after blossom. Look for water-soaked spots on lower leaves and a white fungal growth on the undersides.

e. Mosaic Virus

Affecting a variety of plants including beans, tomatoes and peppers, mosaic virus causes mottled green and yellow foliage or veins. Leaves may curl or wrinkle and plant growth is often stunted. There are no cures for viral diseases such as mosaic.

f. Powdery Mildew

Infected leaves become covered with a white to gray powdery growth, usually on the upper surface. Severely infected plants may turn brown and drop. Fruit ripens prematurely and has poor texture and flavor.

g. Downy Mildew

Downy mildew is a fungal disease caused by Peronospora parasitica. It causes white mold and faint yellow spots on the dorsal and ventral sides of the leaves respectively.

h. Rusts

Often found on mature plants, common rusts affect everything from asparagus to beans, carrots, corn and onions. Infected plants develop reddish-brown spots on leaves and stems.

i. Wilt, Fusarium and Verticillium

Affects a wide range of vegetables including potatoes, tomatoes, peppers and melons. Symptoms include wilting plants and plant parts that turn yellow.

j. Alternaria Leaf Spot

Alternaria leaf spot is caused by fungus *Alternaria brassicae*. The leaves of the infected crop (especially kales) have black or brown circular spots. With time, the spots enlarge and concentric rings appear on them.

k. Frogeye Leaf Spot

Brought about by fungus *Cercospora brassicicola*, Frogeye leaf spot causes pale green, gray or white spots on the leaves. The spots are bordered by a brown ring and can take any shape.

l. White Spot

White spot is a leafy vegetable disease caused by fungus *Pseudocercosporella capsellae*. The crop infected by the fungi usually has brown or gray spots on the leaves. The leaves later turn yellow and fall off within a few days

m. Bacterial Soft Rot

Bacterial soft rot is caused by *Erwinia carotovora*, and it's one of the most common cabbage and collard green diseases. It causes dark, mushy patches on the stems and leaf stalks.

n. Anthracnose

Caused by fungus *Colletotrichum higginisianum*, Anthracnose affects leaves and stems of vegetables like turnips, and causes small, gray or black spots on these parts.

o. Wirestem

Wirestem is caused by fungus *Rhizoctonia solani* which lives in the soil. The disease is recognized by reddish-brown patches on stems of leafy vegetables. The stem rots and peels off with time, exposing a wire-like wood.

p. Bottom Rot

Caused by fungus *Rhizoctonia solani*, Bottom rot makes the lower leaves of the infected vegetables to turn black and wilt. The pathogen can move to the roots, causing root rot.

q. Damping-Off

Damping-off is another leafy vegetable disease caused by *Rhizoctonia solani*. It is also caused by *Fusarium spp or Pythium* spp. The fungus causes decay and wilt in seeds and seedlings respectively. The infected seedlings have light brown stems.

r. Blackleg

Blackleg is brought about by *Leptosphaeria maculans* which is a fungus that causes small, light brown spots on the stems. The spots are usually sunken and can enlarge, making the stem to deteriorate.

s. White Rust

Caused by *Albugo candida*. White rust is a fungal disease recognized by yellow-white spots on the leaves and sometimes on the stems of greens.

t. Yellows

Yellows is a fungal disease caused by *Fusarium oxysporum* con-glutinans, and triggers formation of large, dull yellowish green patches on the leaves of vegetables.

u. Clubroot

Clubroot is brought about by fungus *Plasmodiophora brassicae*, and causes leaves of the infected vegetable crop to turn yellow or pale green. The pathogen spreads to the stem and roots, causing club-like swellings on them.

v. Black Rot

Black rot is caused by both fungi and bacteria (*Xanthromonas campestris*). It brings about yellow patches on the leaves of the infected crop. With time, the patches turn brown. The pathogen spreads to the veins and midribs of leaves and the stem, making them black.

w. Bacterial Diseases

- ✰ Bacterial black spot = bacterial canker: *Xanthomonas campestris* pv. *mangiferaeindicae.*
- ✰ Bacterial fruit rot: *Pectobacterium carotovorum* subsp. *carotovorum* = *Erwinia carotovora* subsp. *carotovora, Erwinia herbicola.*
- ✰ Crown gall: *Agrobacterium tumefaciens.*

ii. Common Diseases of Leafy Vegetables

1. Downy Mildew
2. Alternaria Leaf Spot
3. Frogeye Leaf Spot
4. White Spot
5. Powdery Mildew
6. Anthracnose
7. Wirestem
8. Bottom Rot
9. Damping-Off
10. Blackleg
11. White Rust
12. Bacterial Soft Rot
13. Yellows
14. Clubroot
15. Black Rot
16. Mosaic
17. Root Nematode

iii. Mango

Common Diseases

a) Powdery Mildew (Oidium mangiferae)

The characteristic symptom of the disease is the white superficial powdery fungal growth on leaves, stalks of panicles, flowers and young fruits. The affected flowers and fruits drop pre-maturely reducing the crop load considerably or might even prevent the fruit set. Rains or mists accompanied by cooler nights during flowering are congenial for the disease spread. The fungus parasitizes young tissues of all parts of the inflorescence, leaves and fruits.

Control: Following three sprays of fungicides at 15 days interval recommended for effective control of the disease: Wettable sulphur 0.2 per cent (2 g Sulfex/lit. water). Tridemorph 0.1 per cent (1 ml Calixin/lit. water). Dinocap 0.1 per cent (1 ml/g Karathane/lit. water).

b) Anthracnose (Colletotrichum State of Glomerella cingulata)

The disease causes serious losses to young shoots, flowers and fruits under favourable climatic conditions of high humidity, frequent rains and a temperature of 24-32°C. It is also affects fruits during storage. The disease produces leaf spot, blossom blight, withertip, twig blight and fruit rot symptoms. Tender shoots and foliage are easily affected which ultimately cause 'die back' of young branches. Older twigs may also be infected through wounds which in severe cases may be fatal.

Control: Trees may be sprayed twice with Bavistin (0.1 per cent) at 15 days interval during flowering to control blossom infection. Spraying of copper fungicides (0.3 per cent) is recommended for the control of foliar infection.

c) Die Back (Botryodiplodia theobromae)

The disease is characterized by drying of twigs and branches followed by complete defoliation, which gives the tree an appearance of scorching by fire. The onset of die back becomes evident by discolouration and darkening of the bark. The dark area advances and young green twigs start withering first at the base and then extending outwards along the veins of leaf edges. The affected leaf turns brown and its margins roll upwards. At this stage, the twig or branch dies, shrivels and falls. This may be accompanied by exudation of gum. In old branches, brown streaking of vascular tissue is seen on splitting it longitudinally.

Control: (i) Prune the diseased twigs and spray with copper oxychloride (0.3 per cent) on infected trees. Pruning should be done in such a way that the twigs are removed 2-3 inches below the affected portion. (ii) In small plants, pruning of twigs is followed by pasting of copper oxychloride.

d) Phoma Blight (Phoma glomerata)

The symptoms of the disease are noticeable only on old leaves. Initially, the lesions are angular, minute, irregular, yellow to light brown, scattered over leaf lamina. As the lesions enlarge, their colour changes from brown to cinnamon and they become almost irregular. Fully developed spots are characterized by dark margins and dull grey necrotic centres. In case of severe infection such spots coalesce forming patches measuring 3.5-13 cm in size, resulting in complete withering and defoliation of infected leaves.

Control: The disease could be kept under control by spray of copper oxychloride (0.3 per cent) just after the appearance of the disease and subsequent sprays at 20 day intervals.

e) Bacterial Canker (Xanthomonas campestris pv. mangiferaeindicae)

The disease is found on leaves, petioles, twigs, branches and fruits, initially producing water-soaked lesions and later turning into typical cankers. The water soaked lesions also develop on fruits which later turn dark brown to black. They often burst open, releasing a highly contagious gummy ooze contain bacterial cells. The fresh lesions on branches and twigs are water soaked which later become raised and dark brown in colour with longitudinal cracks but without any ooze.

Control: Three sprays of streptocycline (100 ppm) or Agrimycin-100 (100 ppm) after first visual symptom at 10-days intervals. Monthly sprays of Bavistin (1000 ppm) or copper oxychloride (3000 ppm) were also found effective.

f) Red Rust (Cephaleuros virescens)

Red rust disease, caused by an alga, has been observed in mango growing areas. The algal attack causes reduction in photosynthetic activity and defoliation of leaves thereby lowering vitality of the host plant. The disease can easily be recognized by the rusty red spots mainly on leaves and sometimes on petioles and bark of young twigs and is epiphytic in nature. The spots are greenish grey in colour and velvety in texture. Later, they turn reddish brown.

Control: Two to three sprays of copper oxychloride (0.3 per cent) is effective in controlling the disease.

g) Sooty Mould (Meliola mangiferae)

The disease is common in the orchards where mealy bug, scale insect and hopper are not controlled efficiently. The disease in the field is recognis by the presence of a black velvety coating, *i.e.*, sooty mould on the leaf surface. In severe cases the trees turn completely black due to the presence of mould over the entire surface of twigs and leaves.
Control: Pruning of affected branches and their prompt destruction prevents the spread of the disease, Spraying of 2 per cent starch is found effective,It could also be controlled by spray of Nottasul + Metacin + gumacasea (0.2 per cent + 0.1 per cent + 0.3 per cent).

Mango and its Pests

More than 492 species of insects, 17 species of mites and 26 species of nematodes have been reported to be infesting mango trees, about 45 per cent of which have been reported from India. Almost a dozen of them have been found damaging the crop to a considerable extent causing severe losses and, therefore, may be termed as major pests of mango. These are hopper, mealy bug, inflorescence midge, fruitfly, scale insect, shoot borer, leaf webber and stone weevil. Of these, insects infesting the crop during flowering and fruiting periods cause more severe damage. The insects other than those indicated above are considered as less injurious to mango crop and are placed in the category of minor pests.

iv. Tomato

Common Diseases

a. Early Blight

Symptoms: Early blight affect the foliage, stems and fruit of tomatoes. Dark spots with concentric rings develop on older leaves first. The surrounding leaf area may turn yellow. Affected leaves may die prematurely, exposing the fruits to sun scald.

Management: Early Blight fungus overwinters in plant residue and is soil-borne. It can also come in on transplants. Remove affected plants and thoroughly clean fall garden debris. Wet weather and stressed plants increase the likelihood of attack. Copper and/or sulfur sprays can prevent further development of the fungus.

b. Gray Leaf Spot

Symptoms: Gray Leaf Spot affects only the leaves of tomatoes, starting with the oldest leaves Small, dark spots that can be seen on both the top and bottom surfaces of the leaves. The spots enlarge and turn a grayish brown. Eventually, the centers of the spots crack and fall out. Surrounding leaf areas will turn yellow and the leaves will dry and drop. Fruit production is inhibited.

Management: Warm, moist conditions worsen gray leaf spot problems. Remove all affected plants and fall garden debris. Select resistant varieties.

c. Late Blight

Symptoms: Late blight affects both the leaves and fruit of tomatoes. Late Blight is the disease responsible for the Irish Potato Famine. Late Blight spreads rapidly. Cool, wet weather encourages the development of the fungus. Greasy looking, irregularly shaped gray spots on leaves. A ring of white mold can develop around the spots, especially in wet weather. The spots eventually turn dry and papery. Blackened areas may appear on the stems. The fruit also develops large, irregularly shaped, greasy gray spots.

Management: Copper sprays offer some control. Serenade® works best as a deterrent, rather than a cure. The Late Blight fungus can overwinter in frost-free areas. Since it spreads to potatoes, it also overwinters in potato debris and seed, even in colder areas.

e. Septoria Leaf Spot

Symptoms: The papery patches on the leaves develop tiny, dark specks inside them. Older leaves are affected first.

Management: Copper sprays and Serenade® are somewhat effective at halting the spread of symptoms.

f. Southern Blight

Symptoms: Southern Blight manifests as a white mold growing on the stem near the soil line.

Dark, round spots will appear on the lower stem and both the outer and inner stem will become discolored. Southern Blight fungus girdles the tomato stem and prevents the plant from taking up water and nutrients. Young plants may collapse at the soil line.

Management: Crop rotation seems to help. There has also been some evidence that extra calcium and the use of fertilizers containing ammonium offer some protection.

g. Verticillium Wilt

Symptoms: This name can be misleading, as sometimes the leaves will turn yellow, dry up and never appear to wilt. Verticillium wilt is caused by a soil-borne fungus and it can affect many different vegetables. The fungus can persist in the soil for many years, so crop rotation and selection of resistant varieties is crucial. Wilting during the hottest part of the day and recovering at night, yellowing and eventually browning between the leaf veins starting with the older, lower leaves and discoloration inside the stems. Verticillium Wilt inhibits the plant's ability to take in water and nutrients and will eventually kill the plant. Verticillium wilt is more pronounced in cool weather.

Management: Remove affected plants and choose resistant.

h. Anthracnose

Symptoms: Anthracnose is a very common fungus that causes tomato fruit to rot. Small, round, sunken spots appear on the fruit. The spots will increase in size and darken in the center. Several spots may merge as they enlarge. The fungus is often splashed onto the fruit from the soil. It can also take hold on Early Blight spots or dying leaves. Wet weather encourages the development of Anthracnose. Overripe tomatoes that come in contact with wet soil are especially susceptible.

Management: Copper sprays offer some resistance. Remove the lower 12" of leaves, to avoid contact with the soil. Don't water the leaves, just the base of the plant.

i. Bacterial Speck

Symptoms: Tiny, raised, dark spots, usually with a white border.

Management: Copper fungicide at first signs of symptoms.

j. Blossom End Rot

Symptoms: Dark brown/black spots develop at the blossom end of the fruit and enlarge as the fruit rots. Management: Generally attributed to a lack of calcium during fruit set. This could be caused by too much high nitrogen fertilizer or uneven watering, resulting in fluctuations of nutrient availability.

Management: Remove affected fruit and provide regular, deep waterings.

k. Buckeye Rot

Buckeye Rot is more common in Southern states, especially during wet periods.

Symptoms: Buckeye Rot is similar to Blossom End Rot, except on green fruit. On ripened fruit, the rotting area will appear water-soaked, but not dark in color. The rot develops on the area of the fruit that touches the soil. The spot will enlarge and develop concentric rings that resemble a buckeye. The affected area is smooth, distinguishing it from Late Blight, which has a rough surface.

Management: Remove affected fruit and keep future fruits from contact with the soil.

l. Gray Wall

Symptoms: The green fruits may have a gray cast or gray blotches. Ripe fruit will have green or brown areas on the inside of the fruit.

Management: Good growing conditions will prevent gray wall. Make sure plants aren't heavily shaded, are receiving even waterings and fertilizer and that the soil is not compacted around the roots. Cool temperatures and stressed or unhealthy plants also contribute to the problem.

m. Bacterial Diseases

- ✰ Bacterial Canker of Tomato: *Clavibacter michiganensis* subsp. *michiganensis*
- ✰ Bacterial speck: *Pseudomonas syringae* pv. *tomato*
- ✰ Bacterial spot: *Xanthomonas campestris* pv. *vesicatoria*
- ✰ Bacterial stem rot and fruit rot: *Erwinia carotovora* subsp. *carotovora*
- ✰ Bacterial wilt: *Ralstonia solanacearum*
- ✰ Pith necrosis: *Pseudomonas corrugata*
- ✰ Syringae leaf spot: *Pseudomonas syringae* pv. *syringae*

n. Tomato Mosaic Virus

Mosaic virus attacks many kinds of plants and is common in tomatoes. The tomato plant disease mosaic virus doesn't kill the plant, but it diminishes the number and quality of fruits.The virus gets its name from the markings that resemble a mosaic of light green and yellow on the leaves and mottling on the fruits of affected plants. Leaves may also grow in misshapen forms, resembling ferns. Because the virus must enter through a cut in the plant, avoid handling the plant. Anyone who uses tobacco can easily transmit the disease; wash hands thoroughly with soap to cut the risk of infection.

5.4 Weedicides, Fungicides and Pesticides

Herbicides

Weeds are the unwanted plants which growing with the crop. Herbicides also commonly known as weed killers, are pesticides used to kill unwanted plants and control weeds. Selective herbicides kill specific targets, while leaving the desired crop relatively unharmed. Some of these act by interfering with the growth of the weed and are often synthetic mimics of natural plant hormones. Herbicides used to clear waste ground, industrial sites, railways and railway embankments are not selective and kill all plant material with which they come into contact. Smaller quantities are used in forestry, pasture systems, and management of areas set aside as wildlife habitat. Weedicide are known as the weed killers or pesticides that are used to kill unwanted plants. They are created to kill specific target plants while leaving the rest crop unharmed by its effect. They are widely used in the agricultural practices for better productivity. Weedicides have high water solubility and is persistent in nature.

Common Weedicides

2,4-D Amine Salt 58 per cent SL, 2,4-D Ethyl Ester 20 per cent WP, 2,4-D Ethylester 38 per cent EC, 2,4-D Sodium Salt 80 per cent WP, Ammonium Salt of Glyphosate 71 per cent SG, Anilofos 24 per cent + 2, 4-D Ethyl Ester 32 per cent EC, Anilofos 30 per cent EC, Atrazine 50 per cent WP, Butachlor 5 per cent GR, Chlorimuron Ethyl 25 per cent WP, Fenoxaprop-P-Ethyl 10 per cent EC.

Fungicides

Fungicides are any chemical that can inhibit the growth or development of a fungus. The technical definition: fungicides are any chemical that kills a fungus. *Fungistats* are chemicals that *inhibit*, but do not kill, the fungus. In the turfgrass industry, the term fungicide is commonly used for any chemical that reduces or prevents the development of a fungal disease, but this is much different than the true action of a fungicide. There are many ways chemicals can inhibit or kill a fungus, and there are many different fungi, each of which can react differently to the various fungicides you apply to turf. Further, the effectiveness of a fungicide is determined by much more than just its chemical nature. In short, there is a lot to understand about fungicides if you want to maximize their effectiveness.

Fungicides can either be contact, translaminar or systemic. Contact fungicides are not taken up into the plant tissue and protect only the plant where the spray is deposited. Translaminar fungicides redistribute the fungicide from the upper, sprayed leaf surface to the lower, unsprayed surface. Systemic fungicides are taken up and redistributed through the xylem vessels. Few fungicides move to all parts of a plant. Some are locally systemic, and some move upwardly.

Fungicide residues have been found on food for human consumption, mostly from post-harvest treatments. Some fungicides are dangerous to human health, such as vinclozolin, which has now been removed from use. Ziram is also a fungicide that is thought to be toxic to humans if exposed to chronically. A number of fungicides are also used in human health care. Fungicide, also called Antimycotic, any toxic substance used to kill or inhibit the growth of fungi that either cause economic damage to crop or ornamental plants or endanger the health of domestic animals or humans. Most fungicides are applied as sprays or dusts. Seed fungicides are applied as a protective covering before germination. Systemic fungicides, or chemotherapeutants, are applied to plants, where they become distributed throughout the tissue and act to eradicate existing disease or to protect against possible disease.

Bordeaux mixture, a liquid composed of hydrated lime, copper sulfate, and water, was one of the earliest fungicides. Bordeaux mixture and Burgundy mixture, a similar composition, are still widely used to treat orchard trees. Copper compounds and sulfur have been used on plants separately and as combinations. Synthetic organic compounds are now more commonly used because they give protection and control over many types of fungi and are specialized in application.

Classes of Fungicides, with Examples

Class of Fungicide	*Examples*
Substituted Benzenes	Chloroneb, chlorothalanil, Hexachlorobenzene, pentachloronitrobenzene
Thiocarbamates	Ferbam, metam sodium, thiram, ziram
Ethylene Bis Dithiocarbamates (EBDC's)	Mancozeb, maneb, nabam, zineb
Thiophthalimides	Captan, captafol, folpet
Copper compounds	
Organomercury compounds	Ethyl mercury, methyl mercury, phenyl mercuric acetate
Organotin compounds	Fentin, triphenyl tin
Cadmium compounds	Cadmium chloride, Cadmium succinate
Miscellaneous organic fungicides	Benomyl, cyclohexamide, iprodione, metalaxyl, thiabendazole, triadimefon

Pesticides

Pesticides treat an extensive range of potential plant problems and are applied in numerous ways. Pesticides other than herbicides and fungicides include insecticides, miticides and rodenticides, which kill insects, mites and rodents, respectively. Some pesticides are applied directly to plants or soil and can be in liquid, powder or gas form. Others are baits, such as ant traps or rodent poisons, that appeal to those forms of pests.

Pesticides are chemical substances that are meant to kill pests. In general, a pesticide is a chemical or a biological agent such as a virus, bacterium, antimicrobial, or disinfectant that deters, incapacitates, kills, pests. This use of pesticides is so common that the term pesticide is often treated as synonymous with plant protection product. It is commonly used to eliminate or control a variety of agricultural pests that can damage crops and livestock and reduce farm productivity. The most commonly applied pesticides are insecticides to kill insects, herbicides to kill weeds, rodenticides to kill rodents, and fungicides to control fungi, mold, and mildew.

a. Narrow-Spectrum Pesticides

Pesticides that have a small coverage range are referred to as **narrow-spectrum pesticides**, because they are designed to kill or manage a select group of organisms. Narrow-spectrum pesticides make it possible to target a specific species or group of organisms that are known to cause damage. Many narrow-spectrum pesticides are designed to interact with a characteristic of the pest that is specific to that organism, such as a pheromone, hormone or physical feature. An example of a narrow-spectrum pesticide is chitin inhibitors, which are chemicals that interact with chitin, a component of the exoskeleton of insects. This pesticide inhibits the development of chitin and will eventually result in the death of the insect. The chitin inhibiting pesticide will only harm insects that have chitin in their exoskeletons and will not affect other insects.

b. Broad-Spectrum Pesticides

Although sometimes it is desirable to target a specific species or group of organisms, in some situations, it is necessary to eliminate a wider range of pests that are causing harm. Broad-spectrum pesticides are pesticides that are designed to kill or manage a wide variety of organisms.

Broad-spectrum pesticides are used when many different species of organisms are causing harm or when the specific organism causing harm is unknown. In order to kill or manage such a large variety of organisms, most broad-spectrum pesticides are designed to target a system that is common in many organisms, such as the nervous system or muscular system. An example of a broad-spectrum pesticide is methyl bromide, which is designed to control pests ranging from small insects and pathogens to larger weeds and rodents. The pesticide can be injected into the ground to kill organisms in the soil that might harm the plant while it is growing. It can also be pumped into warehouses or barns to kill pests that could harm the plant during storage or transport for sale.

Types of Pesticides

These are grouped according to the types of pests which they kill:

Grouped by Types of Pests They Kill

1. Insecticides – insects

2. Herbicides – plants
3. Rodenticides – rodents (rats and mice)
4. Bactericides – bacteria
5. Fungicides – fungi
6. Larvicides – larvae

1. Chemically-related Pesticides

a. Organophosphate

Most organophosphates are insecticides, they affect the nervous system by disrupting the enzyme that regulates a neurotransmitter.

b. Carbamate

Similar to the organophosphorus pesticides, the carbamate pesticides also affect the nervous system by disrupting an enzyme that regulates the neurotransmitter. However, the enzyme effects are usually reversible.

c. Organochlorine Insecticides

They were commonly used earlier, but now many countries have been removed Organochlorine insecticides from their market due to their health and environmental effects and their persistence (*e.g.*, DDT, chlordane, and toxaphene).

d. Pyrethroid

The pyrethroid pesticides were developed as a synthetic version of the naturally occurring pesticide pyrethrin, which is found in chrysanthemums. They have been modified to increase their stability in the environment.

e. Sulfonylurea Herbicides

The sulfonylureas herbicides have been commercialized for weed control such as amidosulfuron, azimsulfuron, bensulfuron-methyl, chlorimuron-ethyl, ethoxysulfuron, flazasulfuron, flupyrsulfuron-methyl-sodium, halosulfuron-methyl, imazosulfuron, nicosulfuron, oxasulfuron, primisulfuron-methyl, pyrazosulfuron-ethyl, rimsulfuron, sulfometuron-methyl Sulfosulfuron, terbacil, bispyribac-sodium, cyclosulfamuron, and pyrithiobac-sodium.

2. Biopesticides

The biopesticides are certain types of pesticides derived from such natural materials as animals, plants, bacteria, and certain minerals.

Different Types of Pesticides

1. Bactericides

These destroy, suppress or prevent the spread of bacteria. Examples are swimming pool chemicals containing chlorine, and products used to control black spot (bacterial blight) on garden plants or in orchards. Household disinfectants and some industrial disinfectants are excluded and not considered pesticides.

2. Baits

These may be 'ready to use' products or products which need to be mixed with a food to control a pest. This category includes baits prepared for the control of large animals, such as foxes, wild dogs and rabbits, and baits for insects (such as cockroaches and ants) and molluscs (snail and slug pellets).

3. Fungicides

These control, destroy, make harmless or regulate the effect of a fungus. Examples include chemicals used to treat Grey mould on grape vines and fruit trees, or Downy Mildew on cucumbers.

4. Genetically Modified Organisms (GMOs)

Agricultural crops can be genetically modified to make them more resistant to pests and diseases, or tolerant to certain herbicides. For example, a gene from the bacterium Bacillus thuringiensis can be incorporated into cotton to provide protection against the larval stages of the cotton bollworm and native bollworm.

GMOs are regulated by the Commonwealth Government through the Office of the Gene Technology Regulator (OGTR) under the provisions of the *Gene Technology Act 2000*. Where a genetically modified product is determined to be a pesticide, it is subject to an assessment and registration process in accordance with APVMA requirements.

5. Herbicides

These destroy, suppress or prevent the spread of a weed or other unwanted vegetation, for example, the herbicide glyphosate is used to control a range of weeds in home gardens, bushland and agricultural situations.

6. Insecticides

These destroy, suppress, stupefy, inhibit the feeding of, or prevent infestations or attacks by, an insect. Insecticides are used to control a wide variety of insect pests, including thrips, aphids, moths, fruit flies and locusts. In NSW, pesticides include products used on animals to control external parasites if they require dilution or mixing with water. Products applied directly to animals without dilution, injections or other medicines administered internally to treat animals are veterinary medicines and are regulated by the NSW Department of Primary Industries under the *Stock Medicines Act 1989*.

7. Lures

These are chemicals that attract a pest to a pesticide for the purpose of its destruction. Solely food-based lures, for example cheese in a mousetrap, are excluded and are not considered pesticides.

8. Rodenticides

These are chemicals used specifically for controlling rodents such as mice and rats.

9. Repellents

These repel rather than destroy a pest. Included in this category are personal insect repellents used to repel biting insects.

A number of living organisms that can control pests have also been registered as pesticides. Rabbit Haemorrhagic Disease, for example, has been used to control rabbit numbers; and bacteria that act as biological insecticides have been used to control various insect larvae, such as moths and mosquitoes.

10. 'Natural' Pesticides

Many natural substances can be used as pesticides, such as extracts of pyrethrum, garlic, tea-tree oil and eucalyptus oil. When these natural chemicals are used as pesticides they become subject to the same controls as pesticides produced synthetically.

11. Algaecides

These are used for killing and/or slowing the growth of algae.

12. Antimicrobials

Its control germs and microbes such as bacteria and viruses.

13. Biopesticides

These are made of living things, come from living things, or they are found in nature.

14. Desiccants

These are used to dry up living plant tissues.

15. Defoliants

These cause plants to drop their leaves.

16. Disinfectants

These control germs and microbes such as bacteria and viruses.

17. Miticides

These control mites that feed on plants and animals. Mites are not insects, exactly.

18. Molluscicides

These are designed to control slugs, snails and other molluscs.

19. Mothballs

These are insecticides used to kill fabric pests by fumigation in sealed containers.

20. Ovicides

These are used to control eggs of insects and mites.

21. Pheromones

These are biologically active chemicals used to attract insects or disrupt their mating behavior. The ratio of chemicals in the mixture is often species-specific.

22. Repellents

These are designed to repel unwanted pests

BIBLIOGRAPHY

Textbook

Bean, W.J, Trees and Shrubs Hardy in the British Isles

Cobbett, William. (1833). *The English Gardener.* (Full view). Harvard University.[1]

Compost Everything - composting book by David the Good

Cranshaw, Whitney. *Garden insects of North America: The Ultimate Guide to Backyard Bugs*. Princeton University Press, 2004, ISBN 0-691-09560-4

Damrosch, B. (2008). *The Garden Primer.* (2nd ed). New York: Workman. Dirr, M.A. (1997). *Dirr's hardy trees and shrubs: an illustrated encyclopedia*. Portland: Timber Press.

Damrosch, B. (2008). The Garden Primer. Workman. Retrieved 3 June 2012.

Dirr, M.A. (1998). *Manual of woody landscape plants: their identification, ornamental characteristics, culture, propagation and uses.* Champaign, IL: Stipes. (ISBN 0-87563-800-7)

Dirr, M.A. (2002). *Dirr's Trees and Shrubs for Warm Climates: An Illustrated Encyclopedia*

Di-Sabato-Aust, T. (1998). *The well-tended perennial garden: planting and pruning techniques.* Portland: Timber Press.

Easton, V. (2007). *A pattern garden: the essential elements of garden making.* Portland: Timber Press. (ISBN 0-88192-780-5)

Ely, Helena Rutherfurd (1903). *A Woman's Hardy Garden.*

Ely, Helena Rutherfurd (1905). *Another Hardy Garden Book.*

Ely, Helena Rutherfurd (1911). *The Practical Flower Garden.*

Fisher, S. (1999). *Scented containers: great ideas for year-round fragrance.* NY: Sterling Pub.

Harper, P.J. (2000). *Time-tested plants: thirty years in a four-season garden.* Portland: Timber Press.

Heizer, Roy (2009) Savannah's Garden Plants (Schiffer Publishing)

Howells, J. (1996). *The rose and the clematis as good companions.* Woodbridge: Garden Art Press. (ISBN 1-870673-19-0)

Jarman, Derek and Sooley, Howard. (1995). *Derek Jarman's Garden.* Thames and Hudson.[1]

Lacy, A. (1998).*The inviting garden: gardening for the senses, mind, and spirit.* NY: Henry Holt.

Mabey, Richard. *Flora Britannica.* (1992). Cornell University.[1]

Monty Don's top 10 gardening books." 13 May 2003. Don, Monty. The Guardian UK: Culture–Books.

Morgan, J. Richards, A. and Dowle E. *The New Book of Apples: The Definitive Guide to Over 2,000 Varieties.* (2003). Ebury.[1]

Pears, P. (2002). *Rodale's illustrated encyclopedia of organic gardening.* New York: DK Pub.

Platt, Karen (2004). *Black Magic and Purple Passion: Complete Guide to Dark Plants.*

Rice, B. 2010. Carnivorous Plant Society Archives. The Carnivorous Plant FAQ.

Sakuteiki (11th century) - the oldest work on Japanese gardening

Smith, Jim. The Wise Old Gnome Speaks: How to Really, Really, Really Care About Your Garden

The Gardener's Labyrinth (1577) - noted for its illustrations of Elizabethan gardens

The Hillier Manual of Trees and Shrubs

The profitable arte of gardening (1563) - the first English book on gardening

Williams, B. (1998). *On garden style.* New York: Simon and Schuster. (ISBN 0-684-82605-4)

Websites

http://aggie-horticulture.tamu.edu/ornamental/greenhouse-management/growing-media/

http://aggie-horticulture.tamu.edu/propagation/budding/budding.html

http://agriinfo.in/default.aspx?page=topic and superid=2 and topicid=1009

http://agritech.tnau.ac.in/agriculture/agri_pgr_introduction.html

http://agritech.tnau.ac.in/horticulture/horti_Landscaping_dryflower_arrangement.html

http://agritech.tnau.ac.in/horticulture/horti_Landscaping_garden per cent 20features.html

http://agritech.tnau.ac.in/horticulture/horti_Landscaping_plant per cent 20components.html

http://agritech.tnau.ac.in/org_farm/orgfarm_manure.html

http://agropedia.iitk.ac.in/content/mango-diseases-their-control

http://agropedia.iitk.ac.in/content/manures-and-fertilizers

http://agverra.com/blog/soil-types/

http://byjus.com/bi https://en.wikipedia.org/wiki/Food_storage ology/crop-production/

http://byjus.com/biology/parthenocarpy/

http://byjus.com/biology/plants/

http://byjus.com/chemistry/pesticides/

http://citybugs.tamu.edu/factsheets/ipm/ent-6007/

http://collections.infocollections.org/ukedu/en/d/Jg26p2e/4.5.3.html

http://ecoursesonline.iasri.res.in/mod/page/view.php?id=148108

http://edis.ifas.ufl.edu/pi139

http://encyclopedia2.thefreedictionary.com/Harvesting+of+Crops

http://espacepourlavie.ca/en/importance-and-preparation-soil

http://garden.lovetoknow.com/wiki/Which_Soil_Is_Best_for_Plant_Growth

http://gardeningjones.com/blog/2011/02/27/5-basic-components-of-landscape-design/

http://gardentia.net/manures-and-fertilizers/

http://grounds-mag.com/mag/grounds_maintenance_understanding_fungicides/

http://home.howstuffworks.com/gardening/garden-design/how-to-prepare-soil-for-planting3.htm

http://homeguides.sfgate.com/difference-between-herbicides-fungicides-pesticides-73609.html

http://homeguides.sfgate.com/difference-between-manure-fertilizer-45097.html

http://homeguides.sfgate.com/four-methods-applying-fertilizers-25517.html

http://homeguides.sfgate.com/four-methods-applying-fertilizers-25517.html

http://ipm.ucanr.edu/PMG/diseases/diseases.fruits.html

http://njha.org/contests-projects-activities/horticulture-identification-judging-contest/plant-propagation/

http://oer.nios.ac.in/wiki/index.php/Classification_of_Manures_and_Fertilizers

http://oer.nios.ac.in/wiki/index.php/Introduction_to_Horticulture

http://oer.nios.ac.in/wiki/index.php/Plant_Propagation

http://oer.nios.ac.in/wiki/index.php/Propagation_by_Grafting_and_Budding

http://passel.unl.edu/pages/informationmodule.php?idinformation module=956783940 and topicorder=10 and maxto=10

http://study.com/academy/lesson/manures-fertilizers-types-uses-examples.html

http://study.com/academy/lesson/what-are-pesticides-definition-and-difference-between-narrow-spectrum-broad-spectrum.html

http://timesofindia.indiatimes.com/home/sunday-times/What-is-parthenocarpy/articleshow/3254516.cms

http://vikaspedia.in/agriculture/crop-production/integrated-pest-managment/ipm-for-fruit-crops/ipm-strategies-for-mango/mango-diseases-and-symptoms

http://web-japan.org/kidsweb/virtual/bonsai/bonsai01.html

http://wgbis.ces.iisc.ernet.in/energy/HC270799/HDL/ENV/enven/vol202.htm

http://www.aboutcivil.org/methods-of-irrigation.html

http://www.agriculturalproductsindia.com/fertilizers/fertilizers-chemical-fertilizer.html

http://www.agriinfo.in/default.aspx?page=topic and superid=1 and topicid=356

http://www.aristobiotech.com/products/herbicides

http://www.bestpickreports.com/landscaping/tips/landscape-components-(hardscape-and-softscape)

http://www.beyondpesticides.org/resources/pesticide-gateway/what-is-a-pesticide

http://www.bhg.com/gardening/vegetable/vegetables/tomato-plant-diseases/

http://www.cabi.org/publishing-products/plant-sciences/plant-protection/

http://www.clearias.com/soils-of-india-classification-characteristics/

http://www.cropchemicals.co.in/herbicides-weedicides.php

http://www.cropsreview.com/branches-of-horticulture.html

http://www.cropsreview.com/budding.html

http://www.encyclopedia.com/science-and-technology/biology-and-genetics/environmental-studies/fungicides

http://www.epa.nsw.gov.au/pesticides/pestwhatrhow.html

http://www.epicgardening.com/best-rooting-hormones/

http://www.fao.org/docrep/005/ac681e/ac681e09.htm

http://www.fao.org/docrep/007/y5104e/y5104e08.htm

http://www.fao.org/docrep/t1838e/t1838e0p.htm

http://www.fao.org/docrep/V5030E/V5030E09.htm

http://www.fao.org/docrep/v5030e/V5030E0e.htm

http://www.growthtechnology.com/growtorial/what-is-hydroponic-growing/

http://www.horticultureworld.net/mango-india2.htm

http://www.ikebanahq.org/whatis.php

http://www.importantindia.com/11804/types-of-soils-in-india/

http://www.instructables.com/id/How-to-Make-a-Garland-With-Rose-Petals/

http://www.jains.com/Protected per cent 20Cultivation/greenhouse.htm

http://www.kalyaniind.com/weedicide.html

http://www.landwise.ca/Nutrient/Step4MethodsandTiming.htm

http://www.lwtchort.com/branches-of-horticulture.html

http://www.nytimes.com/1983/07/28/garden/ornamental-topiary-the-green-and-gentle-art.html

http://www.oliviassolutions.com/blog/types-of-plant-propagation/

http://www.preservearticles.com/201012312120/what-is-weeds-and-how-to-control-it.html

http://www.save-on-crafts.com/tecformakgar.html

http://www.sivashakthi.com/plantgrowth.aspx

http://www.superpages.com/em/chemical-fertilizer/

http://www.sustainablebabysteps.com/effects-of-chemical-fertilizers.html

http://www.thebetterindia.com/60350/soil-less-hydroponic-gardening-india/

http://www.theenglishgarden.co.uk/expert-advice/gardeners-tips/how-to-build-a-rockery/

http://www.the-landscape-design

http://www.tomatodirt.com/tomato-diseases.html

http://www.tutorvista.com/content/biology/biology-iv/plant-growth-movements/growth-regulators.php

http://www.wisegeek.com/what-is-fungicide.htm

https://articles.extension.org/pages/64847/what-is-horticulture

https://colourfence.co.uk/blog/the-best-climbing-plants-for-your-garden-fence-or-wall/

https://content.ces.ncsu.edu/plant-propagation-by-layering-instructions-for-the-home-gardener

https://dengarden.com/gardening/Common-Diseases-of-Leafy-Vegetables-Prevention-and-Treatment

https://dengarden.com/gardening/Garden-Components-Used-to-Create-Style

https://dengarden.com/gardening/What-Is-Indoor-Gardening

https://en.wikipedia.org/wiki/Compost

https://en.wikipedia.org/wiki/Conservatory

https://en.wikipedia.org/wiki/Flower_bouquet

https://en.wikipedia.org/wiki/Flower_induction

https://en.wikipedia.org/wiki/Fungicide

https://en.wikipedia.org/wiki/Garden_design

https://en.wikipedia.org/wiki/Harvest

https://en.wikipedia.org/wiki/Ikebana

https://en.wikipedia.org/wiki/Irrigation

https://en.wikipedia.org/wiki/Lawn

https://en.wikipedia.org/wiki/List_of_mango_diseases

https://en.wikipedia.org/wiki/List_of_tomato_diseases

https://en.wikipedia.org/wiki/Manure

https://en.wikipedia.org/wiki/Parthenocarpy

https://en.wikipedia.org/wiki/Peat

https://en.wikipedia.org/wiki/Pesticide

https://en.wikipedia.org/wiki/Soil

https://en.wikipedia.org/wiki/Sphagnum

https://en.wikipedia.org/wiki/Topiary

https://en.wikipedia.org/wiki/Vermicompost

https://extension.illinois.edu/containergardening/herbveggie_containers.cfm

https://extension.umd.edu/growit/food-gardening-101/what-growing-media

https://link.springer.com/chapter/10.1007 per cent 2F978-94-010-1009-2_2

https://link.springer.com/chapter/10.1007 per cent 2F978-94-017-1203-3_17

https://simple.wikipedia.org/wiki/Climbing_plant

https://simple.wikipedia.org/wiki/Peat

https://toxtown.nlm.nih.gov/text_version/chemicals.php?id=23\

https://water.usgs.gov/edu/irmethods.html

https://www.apsnet.org/edcenter/intropp/topics/Pages/Fungicides.aspx

https://www.britannica.com/science/fungicide

https://www.britannica.com/science/parthenocarpy

https://www.cragenomica.es/research-groups/floral-induction-and-development

https://www.epa.gov/recycle/composting-home

https://www.extension.umn.edu/garden/fruit-vegetable/simple-successful/marketing-english/index.html

https://www.ftd.com/blog/design/ikebana

https://www.gardeningknowhow.com/edible/vegetables/tomato/tomato-diseases.htm

https://www.gardeningknowhow.com/garden-how-to/soil-fertilizers/vermiculite-growing-medium.htm

https://www.gardeningknowhow.com/garden-how-to/soil-fertilizers/vermiculite-growing-medium.htm

https://www.gardeningknowhow.com/garden-how-to/soil-fertilizers/what-is-foliar-spray.htm

https://www.gardeningknowhow.com/special/urban/creating-your-own-rooftop-garden.htm

https://www.google.co.in/search?q=green+house and oq=green+house and aqs=chrome.69i57j0l5.5647j0j9 and sourceid=chrome and ie=UTF-8#q=green+house and start=10

https://www.google.co.in/search?q=sphagnum and oq=spha and aqs=chrome.1.69i57j0l5.5071j0j4 and sourceid=chrome and ie=UTF-8

https://www.greenmylife.in/vegetative-propagation-layering/

https://www.landscapingnetwork.com/landscape-design/what-is.html

https://www.maximumyield.com/rooting-hormones-organic-stimulants/2/1480

https://www.ncbi.nlm.nih.gov/pubmed/10637657

https://www.planetnatural.com/vegetable-gardening-guru/plant-diseases/

https://www.plantprotection.org/PlantProtection/Introduction.aspx

https://www.rhs.org.uk/advice/profile?PID=358

https://www.theflowerexpert.com/content/flowerart/driedflowers/making-dried-flower-arrangements

https://www.theknot.com/content/homemade-bouquets-basics

https://www.thespruce.com/garden-design-principles-1403012

https://www.thespruce.com/how-to-preserve-fruits-and-vegetables-1402945

https://www.thespruce.com/how-to-use-rooting-hormone-1902934

https://www.thespruce.com/rooftop-gardening-1403340

https://www.thespruce.com/tomato-leaf-diseases-1403409

https://www.thompson-morgan.com/plants-for-walls-and-fences

https://www.vermiculite.org/resources/vermiculite

INDEX

T

V

W

Z

www.ingramcontent.com/pod-product-compliance
Ingram Content Group UK Ltd.
Pitfield, Milton Keynes, MK11 3LW, UK
UKHW021954270726
14060UKWH00002B/514